# ROBOTS

## WHAT EVERYONE NEEDS TO KNOW®

# ROBOTS
## WHAT EVERYONE NEEDS TO KNOW®

## PHIL HUSBANDS

OXFORD
UNIVERSITY PRESS

# OXFORD
## UNIVERSITY PRESS

Oxford University Press is a department of the University of Oxford. It furthers the University's objective of excellence in research, scholarship, and education by publishing worldwide. Oxford is a registered trade mark of Oxford University Press in the UK and certain other countries.

"What Everyone Needs to Know" is a registered trademark of Oxford University Press.

Published in the United States of America by Oxford University Press 198 Madison Avenue, New York, NY 10016, United States of America.

Library of Congress Cataloging-in-Publication Data
Names: Husbands, Phil, author.
Title: Robots : what everyone needs to know / Phil Husbands.
Description: New York : Oxford University Press, 2022. |
Series: What everyone needs to know |
Includes bibliographical references and index.
Identifiers: LCCN 2021013379 | ISBN 9780198845393 (paperback) |
ISBN 9780198845386 (hardback) | ISBN 9780192584083 |
ISBN 9780192584090 (epub) | ISBN 9780191925948
Subjects: LCSH: Robotics—Popular works. | Robots—Moral and ethical aspects—Popular works. | Robots—Forecasting.
Classification: LCC TJ211.15.H87 2022 | DDC 629.8/92—dc23
LC record available at https://lccn.loc.gov/2021013379

DOI: 10.1093/wentk/9780198845386.001.0001

1 3 5 7 9 8 6 4 2

Printed and bound by
CPI Group (UK) Ltd, Croydon, CR0 4YY

*To CJH*

# CONTENTS

# PREFACE

For some years now there has been a steady stream of far-fetched articles in the media about the impending arrival of super-intelligent robots, along with stories about how most jobs will soon be lost to robots. Many of these myths are whipped up by pronouncements from celebrity technologists (and sometimes scientists) who have little direct knowledge or understanding of AI and robotics research. So it is important to provide a sober, balanced, and informed account of the current state of robotics, where it has come from, and where it might go in the future. The main aim of this book is to explain what robots actually do now and what they might be able to do in the future.

Since the turn of the millennium a quiet revolution has been underway. Millions of autonomous robots with some level of intelligence are now in domestic use, mainly as vacuum cleaners. Driverless cars—which are nothing less than autonomous robots—are starting to appear on our streets, as are delivery and security bots. There is a huge effort underway in industry and universities to develop the next generation of more intelligent, autonomous, mobile robots. Unsurprisingly, these machines, which often seem to display life-like properties, are attracting great attention. Now is the time to learn more about them. How do they work? Do they pose a threat or an unprecedented opportunity?

Even though current robots are nowhere near as intelligent as some would have us believe, and nearly all serious AI and robotics researchers believe the path to major advances is long and difficult, there are still significant ethical issues that arise from today's technology and even bigger ones that will emerge from likely developments in the coming years. There seems to be an irresistible momentum that is bringing us closer to the moment when robots are mobile, widespread and commonly share our environments. But that moment is pregnant with risk. Even barely intelligent machines, or devices that are less intelligent than their makers claim, can cause serious societal issues. Politicians are beginning to wake up to the fact that it is important to understand and plan for these eventualities now. Naturally these issues are of great interest to all of us. So, some of the main problems of robot ethics will be discussed in this book, along with the wider socio-political challenges that robot technology might bring. I question to what extent these concerns are different from those that have arisen in the past from other forms of technology.

Robotics has, and always has had, a complex two-way relationship with popular culture. I try to reflect this throughout the book, and dedicate a chapter to a more detailed discussion of the topic.

But it's not all driverless cars and vacuum cleaners. There are many other interesting, useful, and imaginative robot applications appearing all the time. I try to give a flavor of some of these to give an impression of the ever-growing variety. How these developments come to impact our lives should be largely up to us. That will only happen if we walk into our robot futures with our eyes wide open, so that we are not hoodwinked by duplicitous corporations or shady politicians.

# ACKNOWLEDGMENTS

Thanks to Bart Baddeley, Ken Barrett, Maggie Boden, Josh Bongard, Rodney Brooks, Paul Brown, Andy Clark, Tom Collett, Ezequiel DiPaolo, Alice Eldridge, Dario Floreano, Inman Harvey, Owen Holland, Nick Jakobi, Stefano Nolfi, Michael O'Shea, Rolf Pfeifer, Andy Philippides, Jordan Pollack, Torsten Reil, Anil Seth, Yoonsik Shim, Mike Wheeler, the late John Holland, and the late Jack Good, among many others, for helpful discussion over the years on topics related to those covered in this book.

Thanks to Lewis, Ella, and Theo for sci-fi conversations, to Lewis for help with figures, and to Alison for help with illustrations, proofreading, and much else.

Thanks to Natasha Walter for access to material relating to Grey Walter.

Many thanks to the team at OUP. To Latha Menon for suggesting, encouraging, and organizing this project, as well as for insightful editorial comments, and to Jenny Nugée for all her help. Thanks to the anonymous reader for very useful comments on an earlier draft of the book. Also many thanks to Felshiya Samuel and the team at Newgen for seamlessly managing the production stages.

# 1

# ROBOTS ARE HERE

*What do we think about when we think of robots?*

The most common idea of a robot has changed little in nearly a century: some kind of mechanical man or woman—humanoid machines capable of performing many of the tasks we engage in all the time, such as walking, talking, picking things up and moving them around, as well as some of those that most of us try to avoid, such as indiscriminate acts of death and destruction. Later we will see that this image—and indeed the very idea of a robot—comes from the world of fiction. While it is true that these myths and dreams have seeped into the collective conscious and undoubtedly influence some of the scientific work in the field of robotics, the current reality—though full of enormous interest and potential—is a little less dramatic. As these more mundane machines become commonplace we will start to think of robots in a very different way.

*Why should we care about robots?*

Robots have been used in some industries for more than half a century, so why should we suddenly care? Twentieth-century industrial robots—the kind found on car production lines—mainly operated out of the public gaze and, although sophisticated, were essentially dumb, so they never roused much debate. But since the turn of the millennium new robotic breeds have started to appear at an accelerating pace. Robots that show glimmers of life-like intelligence have started to appear in our homes (autonomous vacuum cleaners) and on our streets (driverless cars), and are being developed for all kinds of applications—some benign, some troubling. It is these mobile, intelligent robots—on the cusp of becoming integrated into our everyday lives—that are now sparking great interest. What can they actually do now and what might they be able to do in the future? Will they force us to change the ways we think about technology and the uses of technology? Will they fundamentally change the ways we live and work?

I will try and answer these and many other questions in this book, but the first point to stress is that it is not a matter of when or if the robots come: they are already here. They are not completely ubiquitous yet or particularly smart or threatening, but, just like computers, the internet, and networked mobile devices before them, there appears to be such momentum behind the development of robot technology—which of course builds on all those other technologies just mentioned—that its spread is inevitable. All over the world, government and corporate R&D funds are being poured into AI and robotics. That is why it is important to have some kind of understanding of the state-of-the-art of these machines and to think about how they should be used and regulated now and in the future. Mobile, physical, autonomous devices that share our domestic and working lives could potentially have a much bigger impact on the world than their predecessor technologies. In the future they could be seen as the crowning glory of twenty-first-century applied science and engineering, or as tools of social inequality and economic oppression. The development of improved robot technology may be inevitable, but the way in which we apply it is not.

Cities such as San Francisco are already the canaries in the mine as far as learning to live with, and legislate for, robots is concerned. At the end of 2017 the San Francisco Society for the Prevention of Cruelty to Animals (SFSPCA) caused a wave of headlines after a security robot it was using to patrol its campus and surrounding public footpaths was attacked.[1] SFSPCA deployed the robot to discourage inhabitants of nearby tent encampments of homeless people, some of whom they suspected of antisocial behavior. The robot was equipped with multiple sensors, including video cameras and face recognition systems. In many ways it was a mobile surveillance unit. As well as incensing the homeless people, who found themselves monitored and recorded, somewhat ironically the 180-kg robot annoyed a local dogwalker because the machine upset her animal, who perceived it as a threat. Within a few days of starting its trundling patrols on public paths, the robot was ambushed. It had a tarpaulin flung over it, was knocked down, and had barbeque sauce smeared across its sensors. Such robots had operated successfully enough in enclosed privately owned spaces such as shopping malls, casinos, and car parks. Bringing them out into the open had caused the problem. This and similar incidents have rushed city authorities in the USA and elsewhere into working out how they should regulate the use of robots in public areas.

This incident tells us two things. First, anyone who thinks introducing robots into the public arena will be straightforward had better think again. Second, it's not difficult to disable most of the current generation of robots.

In San Francisco it was quickly decided that it was not permissible to operate security robots on public streets without special approval. At the same time, the city was finalizing legislation that would allow local delivery robots to travel along footpaths, at least for six-month trial periods.[2] It might be thought this application was likely to be perceived as less of a threat, but some residents were concerned about their footpaths becoming cluttered with autonomous delivery carts, and worried about how safe the technology was, not to mention the impact on delivery jobs. As robots proliferate, so will such understandable concerns. It is up to our politicians to legislate wisely, and for that they, and all of us, need to know what we are getting into.

We are used to machinery that we control. Autonomous mobile physical devices (machines that control themselves) are another story—one that we have little experience of—so we need to proceed with care.

### Why do robots fascinate and sometimes scare us?

Most of us are interested, at least to some degree, in the latest flashy technology. Certain types of robots penetrate deeper.

We are biologically hardwired to react strongly to other autonomous creatures,[3] particularly when they move, because they might represent danger or help or safety. Our response evolved in relation to humans and animals, but it is not surprising that it generalizes to artificial creatures. Autonomous mobile robots arouse unconscious, ancient instincts, which partly accounts for why they fascinate us and have been a major trope of dystopian fiction for more than a century, and why robot scare stories are always popular with the media.

In the mid-1990s, when it was very rare for autonomous robots to be sighted outside research labs, I brought a fairly large eight-legged walking robot home on the way to an international conference where it was to be demonstrated. Some PhD students and I were doing last-minute checks and refining some of the control software. Suddenly my young children and some of their friends and a few stray adults burst into the room on their return from school. It seemed like an ideal opportunity to see how well the robot would perform in a crowded environment full of fast-moving noisy obstacles. One of the research students put the (bright purple) creature on the floor and turned it on. Its behavior was to move around as efficiently as possible using its sensors to avoid bumping into anything. Some of the children backed off warily and ran out the room, some got down on the floor and excitedly tried to interact with the robot, laughing with delight as it moved away or scurried off at high speed into an open space. Most of the adults, apart from the students, hung back, some turning pale.

If the robot had looked like a wheeled remote-control toy car it's unlikely anyone would have batted an eyelid. But it didn't. Its body was reminiscent of a giant insect and it moved its legs in an uncannily natural way. The artificial nervous system we had created for it was strongly inspired by insect neurophysiology and its behavior was deliberately created to appear as life-like as possible. What's more, there was no one stood about with a remote radio control unit. The robot's own on-board nervous system generated its behaviors; once the on-switch had been flipped it was on its own. It was autonomous. It was as if some kind of artificial animal or alien species was marching around the living room floor.

These life-like attributes spoke directly to the deep animal instincts in most of those present. We cannot help but react strongly to something that appears alive, whether with fascination or horror.

Similar forces are at play when it comes to humanoid robots,[4] but with further powerful layers. As a social species we are profoundly interested in ourselves and those like us. Because in our evolutionary history other humans were likely to be our most helpful allies but also among our most dangerous foes, we have many complex instinctual behaviors that surface when we interact with other people. It is no surprise that we are often naturally drawn into interactions with friendly, humanoid robots,[5] and recoil from threatening ones[6]—themes that have long played out in science fiction. As we will see later, the first of these tendencies can be very useful in certain helpful therapeutic or educational settings. The second is ripe for malicious exploitation.

Interestingly, there is some evidence that our empathetic reactions to human-like robots are strongest when their resemblance to real humans is either quite high (but far from perfect) or extremely high (near perfect). There is an area in between, dubbed the uncanny valley by Masahiro Mori, where the resemblance is strong but not perfect. Many of us perceive robots in the uncanny valley as somehow weird or creepy: something is not quite right.[7] Mori compared the effect to viewing a dead body.

It is worth bearing these primal reactions in mind when we explore what robots can and can't do. It is easy to be fooled into believing a robot is much smarter or more capable than it actually is, just because it looks human or appears to behave in a particularly natural way.

### How many robots are out there now?

Far more than most people realize. It is difficult to calculate exact numbers, but a reasonable estimate is that in 2019 there were 13–15 million robots in regular use around the globe. They are being used in industry and commerce, homes, hospitals, schools, and for much else besides.

The numbers break down roughly as follows. The International Federation of Robotics estimates that in 2019 there were 2.6 million industrial robots in use.[8] Those are the heavy-duty machines, usually giant arms, that weld, spray paint, pick, move, assemble, and so on in various manufacturing industries. At the moment, by far the biggest robotic population is robot vacuum cleaners. By 2018 the world leader in this sector, iRobot, had sold more than 20 million robots.[9] Not all of those will still be in use and there are other providers selling large numbers of machines, so a conservative estimate, taking into account global sales for this sector of about 7 million robots a year, might be 10–12 million robot vacuum cleaners in use. Various sources suggest that there are probably about 0.5 million specialist service robots in use,[10] possibly far more. This area includes security robots, educational and home-help robots, telepresence robots, medical robots, bomb-disposal robots, farming robots, and autonomous vehicles. It is the most difficult area to estimate numbers for, and is a sector that is likely to see explosive growth over the coming years.

As might be expected, the use of robots varies from country to country. South Korea, Japan, and Germany have among the highest densities of robots in their manufacturing industries, while the UK has among the lowest in Europe. China and the USA are large markets for industrial robots, with Japan the biggest supplier of this technology. Meanwhile, in many areas of the nascent service robotics sector the USA leads the way in developing and embracing these machines, while Japan has long been the trailblazer on such advanced applications as home-care robots.

These numbers do not include the many tens of thousands of research robots in university and industry labs, the numerous build-your-own kits in school technology classrooms, and the vast numbers of machines designed and built by hobbyists and gadgeteers.

Robots are here; they are proliferating and developing. They are still invisible to many—maybe most—but, at the current rate of change, that won't be the case for much longer.

# 2

# THE BASICS

*What exactly is a robot?*

In comparison to the standard Hollywood fantasy, the robotics community has rather more general, and often slightly humdrum, ideas of what a robot is. A typical working definition goes something like this: a physical device capable of autonomous or pre-programmed behavior in the world involving interactions with its environment through sensors and actuators (components, such as motors, that cause some part of the machine to operate or move)—so a machine that uses sensors to gather information about the world and actuators to enable it to perform actions in the world based on some processing of the information it has sensed.

This is a very general definition, which on the face of it might include a simple security device that uses motion-detecting sensors to trigger a circuit that turns on a powerful spotlight. Or a home virtual assistant that uses microphones to listen out for commands and electromagnets to drive a speaker cone which pumps out sound waves carrying its replies. But we wouldn't normally think of either of those devices as robots. One is just a fancy light switch and the other is an interface to powerful cloud computing services.

The truth is that I have never come across a general definition of a robot that, when examined closely, doesn't include very simple devices such as thermostats, which sense the temperature and turn the heating on or off depending on whether it is too cold or too hot. Any definition will always encompass a continuum of machines. The kinds of machines I will be referring to as robots are physical devices that sense the world, act in the world, and through those actions change the world; in most cases this will involve movement—movements of the entire robot or parts of the robot, and often movements of external objects out in the world. Think of a wheeled mobile robot (that might look a bit like R2-D2 from *Star Wars*) roaming around exploring its environment, or an industrial robot arm moving objects around on a production line.

### What different kinds of robots are there?

There are many different types of robots (Figure 2.1), and by breaking them down into categories we can start to get a more detailed picture of what the possibilities are.

Let's start with the basic body types. There are *wheeled robots* which have a wheel-driven base on top of which all manner of structures can be built. These include small disc-like robot vacuum cleaners, warehouse robots, hulking great robot security guards, and also driverless vehicles—from cars to large-goods vehicles to farm tractors. Robots that are driven by caterpillar tracks are closely related. Tracks handle certain types of outdoor terrain a bit better than wheels, particularly soft surfaces, and are used on some types of bomb-disposal robots, some agricultural robots, and some military robots.

*Legged robots* locomote using two or more legs (often four, six, or eight), although there have been one-legged hopping robots too. Legs are much more difficult to control and coordinate than wheels, but they have some distinct advantages in certain environments. They are much better at dealing with rugged, difficult terrain and allow flexible stabilization and adjustment to the body position (e.g., keeping it clear of dangerous rocky bumps). Sophisticated multi-jointed legs can also

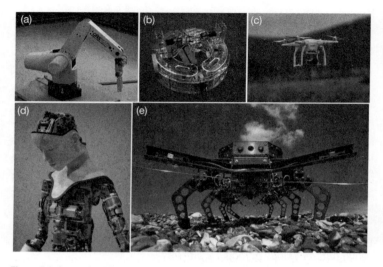

**Figure 2.1** Some robot types. Clockwise from top left: (a) industrial arm, (b) wheeled, (c) flying, (d) legged, (e) humanoid.

*Source:* a and b: Phil Husbands; c: Jared Brashier/unsplash; d: © Roger Bamber; e: Franck V/ unsplash.

be used to manipulate the world—for instance to clear obstacles out of the way. By changing gaits, a legged robot can smoothly and rapidly switch between slow, careful movements, brisk walking, trotting, or running.

*Humanoid robots* are shaped like us, with a head, a torso, two arms, and usually two legs (although some are fixed to bench tops, lacking legs and a lower body). Many advances in this area have come from Japan. The head will often be able to rotate up and down and from side to side, and contains a variety of sensors including cameras and microphones acting as eyes and ears. Some humanoid robots have complex jointed arms, legs, hands, and feet, while others employ simplified limbs and graspers. Some have bodies that are quite abstract representations of humans, while others are eerily accurate. Realistic faces that can be moved by motors and artificial muscles to form various different expressions are becoming more common, such as those found on the ultra-realistic-looking humanoid robots pioneered by Japanese roboticist Hiroshi Ishiguro.

Humanoid robots come in all sizes, from 20 cm to more than 2 m tall. One of the main drives for developing such machines is the belief that they should be able to blend into our lives more easily than other types of robots. On the whole, we find ourselves drawn into natural, instinctual interactions with humanoid robots—they tap directly into our innate anthropomorphism. Robots that move in the same way and are the same size as us should be able to share our domestic arrangements without major alterations to our homes. Robots that can walk up and down stairs, reach into cupboards, and easily pick their way around the mess and jumble on our children's bedroom floors could make the best home-help robots, or so the argument goes.[1]

Increasingly popular are *flying robots*. These are most likely to be some kind of drone, or unmanned aerial vehicle (UAV) to give them their more technical name, often with additional sensors and on-board computers. There are also more exotic animal-like flying machines based on insect or bird mechanisms that involve flapping wings. The military have a strong interest in UAVs. As well as being employed in security and surveillance applications, they are commonly used in the entertainment and artistic industry as a convenient way of gathering aerial footage or creating airborne displays.

More specialized are *swimming or underwater robots*. Some of these are unmanned miniature submarines used for maintenance and inspection of undersea constructions, while others are shaped like fish and use more biologically inspired locomotion: they swim. There is growing interest in the development of fish-like swimming robots as they are more maneuverable than standard submarines and can potentially move in a more energy-efficient way. These can be important considerations when inspecting complex man-made structures or studying the state of areas of the ocean bed or the local ecology in parts of the oceans. Realistic-looking robot fish can also be used to study the behavior of real fish, blending in as they act as unobtrusive monitoring devices.[2] Very soon we may also start to see *robot ships*. The US navy is exploring such technology, and experimental driverless container ships are already being tested on some of the world's busiest shipping lanes.[3]

And of course there are complex articulated *robot arms*, of the kind that are widely used in industry.

Most of the robot body types mentioned are built from hard, rigid structures and casings. Some robot fish are the exception: they have soft, flexible bodies or body parts. Many animals, including us, have largely soft bodies, even if they are supported by a rigid internal skeleton. Soft bodies come with a number of interesting advantages. Squishy, highly compliant materials bring with them increased flexibility and adaptability. If the body can be flexed and distorted into numerous configurations, it may be much easier to squeeze through tight openings and navigate around difficult environments. (Think of a robot octopus squashing through narrow gaps and pulling itself along partially blocked passages as it searches a submerged wreck, while a traditional rigid submarine gets stuck at the entrance.) Soft bodies can also provide highly energy-efficient locomotion. Hence the field of *soft robotics* has recently taken off. Biologically inspired, soft-bodied robots can also be much safer for interactions with humans. Being accidently clunked with a spongy, compliant arm is likely to be much less painful than being hit with a solid steel limb. There is a lot of ongoing work on soft robotic actuators for grabbing and manipulating objects—typically they use much less force than traditional rigid actuators, and can delicately manipulate fragile items. They are often made from new kinds of shape-shifting materials. This is a new but very exciting area of robotics that promises much for the future.[4] Artificial muscle, sinew, and flesh could also lead to more and more realistic humanoid or animal-like bodies.

## Do they all behave in the same way?

No, far from it. To answer in more detail let's turn to the types of behaviors robots can exhibit. Irrespective of what kind of body a robot has, or how it gets around, there are some useful distinctions we can make about behavior. At one end of the scale is *remote-control* behavior. The robot is simply dancing to the tune of a human puppet master. All its movements and responses are down to the skill of the human operating the robot's actuators via a remote-control unit. Bomb-disposal and military robots often fall into this class, as do medical robots such as those used in surgery, many types of flying drones, and telepresence robots which allow people to have an interactive virtual presence at a particular location. A typical example of the latter is an executive at a board meeting who can move around or turn to face a presentation or another attendee via a robotic system, with cameras, microphones, and speakers all hooked up to video conferencing, and whose actuators can be controlled over the internet. Another example of robotic tele-presence is a doctor interacting with a distant patient via re-motely moveable cameras and a screen.[5]

One rung above this, from the robot's perspective, is *inflex-ibly dumb pre-programmed* behavior. In this case the robot con-trol system is in charge. The robot typically performs the same action over and over again according to a rigid protocol from which it can never deviate. Certain types of manufacturing robot fall into this class. Interaction with the environment and use of sensory data is usually very limited. An example is a robot arm (Figure 2.1) working on a production line that is pro-grammed to pick up component A from location X, move it to location Y, where it is then pushed onto component B, and then repeat ad nauseam. If the wrong component is at X, or Y is blocked, the robot has no sensible adaptive response; it does not "know" what to do when things go wrong.

Because industrial robots are usually programmed to re-peat precise manipulations, such as welding or paint spraying, their working environment is often carefully designed so that intricate sensory feedback is unnecessary—the robot performs its repetitive tasks in an accurate, efficient, but essentially unintelligent way, without having to take much notice of the world around it. Although they make use of sensors to help control and fine tune their movements, and to avoid major collisions, the behavior of such robots cannot be said to be in any way intelligent—if the car on the production line is misaligned or is not the expected shape, the robot cannot react to the new set of circumstances, it can only repeat its preprogrammed movements. While these limitations are manageable in the highly controlled environment of a production line, they become problematic if a robot is to be used in less-structured and less-predictable circumstances.[6] More complex cases involving cluttered or dynamic or noisy environments, or delicate manipulations of objects, usually require more sophisticated sensory feedback and perceptual processing—for instance the use of a vision system to help guide a robot to the desired location. This brings us into the realm of AI and *intelligent, adaptive robotics*.

The need for intelligent, adaptive behavior is even more obvious when robots break out of the confines of a production line and become mobile. A robot involved in an exploration or search-and-rescue mission must interact with an environment it has not experienced before and somehow cope with unexpected scenarios and uncertain outcomes. The control methods used for industrial arms are no longer sufficient.

In contrast to dumb, inflexible machines, *autonomous* robots are required to behave in an appropriate way in whatever circumstances they find themselves. Like biological creatures, their behaviors must be self-generated, making use of sensory information to moderate their responses to the world. No one else is holding the strings; they must make their own decisions, fashion their own actions. Mobile autonomous robots, which while roving around are likely to come across a range of unpredictable and unforeseen situations, are the most interesting and the most challenging: challenging from a technical point of view—how do we make them adaptive and intelligent enough to cope?—and from a societal standpoint—how do we integrate them into our lives? This is the type of robot that will most concern legislators and technologists as the twenty-first century gathers momentum, and will be a major focus of this book.

All of which brings us to the next big question.

### How do robots work?

I'm not going to go into much detail yet—that will wait for later (see Chapter 4)—but at this stage it is important to give a broad-brush account of how robots function. At the very highest level, Figure 2.2 captures what is common to all robots. The gray area indicates the components that are part of the robot. The robot interacts with the world: its environment. Sensors, the numbers and types of which can vary wildly from robot to robot, capture some kind of information about the environment, the external world. The sensor signals are acted upon in some way by the robot control system, perhaps coordinating information from multiple sensors, perhaps filtering or transforming the data streaming in through the sensors. Control system designs and operational principles can differ enormously between different types of robots. The

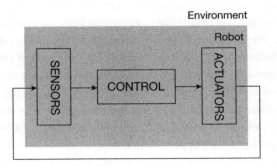

**Figure 2.2** The robot sense–act loop.

control system sends commands to the robot's actuators—its motors or artificial muscles or grippers or whatever array of mechanisms it has to enable movement and changes to the world: to initiate actions. The controller coordinates and modulates information flow between the sensors and actuators, and hence between sensing and action.[7] Notice the arrow in Figure 2.2 that goes from the actuators through the environment and back to the sensors (a sense–act loop). This reminds us that there is always feedback between sensing and action: the two are often tightly bound together. When an action causes a movement in the world, or a change to the world, the sensor reading will very likely change. Sensing enables action and action sets up further sensing. In nature, a hunting animal spies its prey, moves its head to track it, and then runs forward to get a closer view; sensing and action are inextricably linked.

Let's get a bit more concrete by examining a very simple example. Consider the basic wheeled robot shown in Figure 2.3. It has motor-driven front wheels and a free wheel at the back for stability. The front wheels can be independently driven at various speeds forward and backward by their motors. Different combinations of speeds for the two wheels allow the robot to move in various ways. It can move in straight lines (forward and backward), steer to the left or right (in gentle and tight curves), or spin on the spot.[8]

We can see from Figure 2.3 that the robot has a number of short-range infrared (IR) proximity sensors evenly spaced around the bottom edge of its body. These can detect if there is an object nearby. These work by the emitter part of the sensor firing out IR beams which are reflected back off nearby surfaces to be picked up by the receiver part of the sensor. A high sensor reading means that there is something nearby in the line of sight of that sensor. IR is just a particular frequency of electromagnetic radiation, typically used for TV remote control.

Let's put this robot in the same scenario as the more complex eight-legged robot from the last chapter—the one that delighted most of the children and frightened the adults—and expect it to exhibit the same basic behavior: move around a room avoiding obstacles.

A simple control system that would enable the robot to do this to some extent would work as follows and could be implemented in software running on an on-board computer chip in the control unit (Figure 2.3). The controller repeatedly reads all the IR sensors; if any of the sensors on the right side return a high reading there must be an object in the vicinity on the right, so the motors are activated so as to turn the robot to the left (the right motor is driven faster than the left). The higher the reading the sharper the turn, which should move

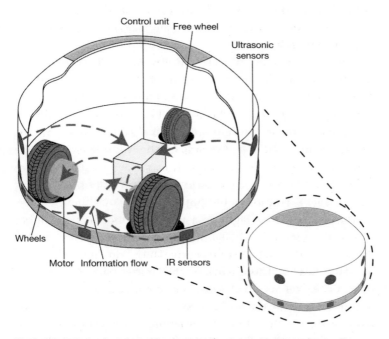

**Figure 2.3** A simple wheeled robot. Foreground: external view with sensors showing. The blow-up, cut-away view shows a schematic representing the internal workings. Information flows from the sensors to the controller and from the controller to the motors.

*Source:* Lewis Husbands.

the robot away from the obstacle. A similar thing is done for the left side, and if no sensors are high the robot must be away from any objects so it moves forward in a straight line by driving both motors at the same speed.

This basic control system would work reasonably well in a fairly uncluttered environment but could soon get into trouble in a crowded area. If the robot drove into a corner, or a narrow gap behind a piece of furniture, or if it was heading perpendicularly toward a wall, it would get stuck. The left and right front sensors would both be triggered, perhaps one side first and then the other, and then both together. The robot would steer first one way and then the other, jiggling back and forth in a demented dance until it hit something and got jammed. To overcome these limitations extra conditions could be added to the control system. When it got stuck, perhaps the robot should reverse and move in the opposite direction from which it had just come and then turn by a random amount before driving off again.[9] Similarly, when sensors on both sides are triggered at the same time, perhaps the robot should take evasive action by again backing away and heading off in a random direction. This second version would be less likely to get stuck but will still be far from perfect.

The point is that even in this very simple example you can start to see that the job of the controller—to coordinate actions in response to sensor signals so as to produce a particular behavior in the world—can soon become quite tricky. As we will discuss in later chapters, there are many types of control architecture and control principles that are used, ranging from simple rule-based systems like the one just described (if this happens, do such and such; otherwise, if that is triggered, do the following; and so on) to artificial neural networks. The latter are inspired by biological nervous systems; parallel signals flow around distributed networks of simple processors that work a little like neurons.

### Can different sensors change a robot's capabilities?

Yes, and often dramatically. If we used the same robot described above (Figure 2.3) in the same scenario, but with different sensors, the story could be radically altered. The simple IR sensors mentioned so far can only return information about things very nearby (so-called proximal information), and so the robot can only react when it is already very close to hitting something. But what if the robot had sensors that returned information about objects in the distance (so-called distal information)? One such sensor is a laser range-finder which measures the distance to an object from the time it takes for a pulse of laser light to travel from the sensor to the object and back again after reflection.[10] Typically, the ones used on robots have an accuracy of 1–2 cm and a range of up to 30 or 40 m. Ultrasonic sensors (like those on the robot in Figure 2.3) work in the same way, except using high-frequency sound waves; their range is usually lower (about 10 m). If the robot made use of a ring of such sensors, instead of just the IR sensors, it would have continual access to more or less accurate information about the distances and bearings to all objects in the room, or at least those within the line of sight of the rangefinders (they can't detect things hidden behind other larger items).

In order to make good use of this extra information, the control system would have to grow in complexity from the simple one sketched out above. We would want it to use this information to plot a sensible course around the room, now aware of potential hazards long before it reaches them. The robot could then accurately and smoothly vary its speed and change its direction according to how close it was to obstacles. We might even want it to make a few exploratory laps of the room, scanning as it goes, and build up a kind of map of where objects are. It could then use this map to plot its course. This is precisely what is done in many commercial applications of robotics.

Another very interesting type of proximal sensing employed by some robots is vision. Vision, via camera sensors, potentially carries a huge amount of information. High-resolution color vision opens up access to all kinds of details that might enable the robot to identify what the object *is*, rather than just the fact that it is there. A home-help robot might be able to use vision to identify a particular food their owner has asked for (an apple rather than a pear, say) or to individually recognize the various inhabitants of the house. A security robot might use vision to spot an intruder. An industrial robot might use vision to help guide a delicate assembly job. It is no coincidence that vision is so widely used in the natural world; it is an incredibly useful sense. However, biological visual systems are highly complex (a large chunk of our own brains is dedicated to vision[11]), and artificial visual systems, as used in robotics and AI, are a long way behind. Designing control systems that can reliably process detailed visual information is far from easy. All of the robotic visual tasks listed earlier in this paragraph can be achieved to some degree, but usually only if the lighting is just so and the subject is facing the camera at the right angle. Compare that to our own vision which copes with wildly varying lighting conditions, shadows, and partially obscured images, and somehow allows us to recognize complex objects (such as other humans) from almost any angle. However, progress is being made, and less-detailed, low-resolution vision—which is generally easier to handle—is very useful in some applications.

### How do legged robots walk?

The demands on the control system are also increased when a more complex array of actuators is involved. Let's return to the simple scenario of the robot moving around a room while avoiding obstacles, but this time consider a robot with many legs and joints that need controlling. Imagine a robot with eight legs (like the one in Figure 2.1), each with a hip joint and a knee joint, arranged around its body a bit like a spider. Let's say the hip joint can move up and down and side to side. That requires a minimum of two actuators, maybe two motors—one for each direction of rotation (known as degrees of freedom). Suppose the knee just moves up and down, requiring one actuator. That is a total of three actuators per leg or twenty-four in total. Now the control system has to coordinate the activity of twenty-four actuators instead of the two for the simple wheeled robot. In order to make the robot move forward or backward, fast or slow, or rotate on the spot, or steer to the left, *all* those legs have to be carefully choreographed; otherwise the robot will crash to the ground, thrashing around in desperation.

Actually any decent-quality legged robot will involve multiple sensors on each leg, recording limb angles and velocities and so on—information that can come in very handy for coordinating movement—so the control system for such a robot will also have to handle far more sensors as well as actuators. Again, many different kinds of control architectures and principles can be used in this case. Complex rule-based systems can be made to work, specifying the relative timings and direction of movements of the legs, but tend to result in very clunky, juddering movements, whereas distributed artificial neural network controllers, based on the architecture of biological nervous systems, tend to produce much more fluid, natural-looking motion.[12]

When it comes to walking robots, the nature of the body is also a big factor in how difficult the control task is. An eight-legged spider-like robot, although there is much to coordinate, at least has a big advantage when it comes to stability. It is not too difficult to maintain a relatively stable gait as long as a certain number of legs are firmly on the ground at any moment—say half of them, but distributed roughly evenly around the body, not all on the same side or it will just keel over! But a two-legged humanoid robot is quite another story. Bipedal locomotion, as biologists call two-legged walking (or running), is inherently unstable. Our own mode of walking is a kind of controlled falling, which is why we have such complex meshes of leg and lower back muscles, and why we immediately collapse if they stop working. This instability is also why some humanoid robots walk with an unhuman-like, rather undignified, crouched posture as they frantically try to remain balanced.[13]

### Is there one best approach to robot design?

It will be obvious by now that there is no simple answer to the question "how do robots work?" There are many ways to make a robot work, all depending on the exact relationships between the sensors, the body, the control system, the actuators, and the environment. Variations in any of those can make big changes to the robot's capabilities and the ways in which the other parts of the system must operate. Finding the right balance between all these factors to produce the desired behavior is at the heart of the art of robot design. As the required behavior becomes more demanding, usually because it entails some degree of adaptive, intelligent action, the design task becomes considerably harder. We'll discover later that we don't have to go very far up the scale of "intelligence" before we are out of our depth and do not know how to make the robot produce the desired behavior. We have a long way to go before we can produce anything like natural, biological intelligence—which has to deal with a wide range of often unfamiliar circumstances. But within narrow confines, impressive progress has been made. Robots that can recognize objects, make decisions, and adapt to changing circumstances do exist, but their range of behavior is limited and their modes of operation tightly restricted. Generally they are specialized to a single task.

More complex behaviors require more sophisticated control, and robot designers often look to biology for inspiration; after all, the only examples of general, adaptive intelligence we have are all from the natural world. One interesting lesson from biology is that in most animals the boundaries between the body, the sensors, the actuators (muscles), and the control system (the central nervous system) are very blurred. For instance, our retina, which might naïvely be thought of as equivalent to the sensor chip in a digital camera, because it lines the back of the eye, is actually brain tissue and is normally classified as part of the nervous system. Recently there has even been a challenge to the idea that the central nervous system does all the information processing. It has been shown that networks of muscles are capable of processing sensory information and could play a key part in generating and controlling motor behaviors. It is not just the brain that coordinates and controls, but the body too.[14]

Related lessons are being learnt about the importance of the overall morphology of an autonomous agent—biological or artificial—in terms of the body shape and sensor layouts and the way the actuators integrate with the body, and the whole ensemble interacts with the environment.[15] So, for instance, the kind of panoramic sensing often found in the animal kingdom—such as being able to see in all directions all the time—can make many tasks easier. Although it may at first seem like a complication that just increases the amount of processing the brain must do, 360-degree panoramic vision can be very helpful in avoiding predators or spotting prey, and for tasks such as navigation. Being able to see the way the world changes in all directions as an animal moves provides very rich information for trying to pinpoint a location. This is partly why many insects, even though they have limited neural resources, have panoramic vision. On the motor side of things, the exact shape and relative dimensions of limbs and joints, coupled with the properties of tendons and ligaments, can lead to highly efficient movements which exploit gravity and conserve energy.

Trying to understand the harmonious integration of brain and body helps us to both uncover the way behavior is generated in nature and design better robots. It is a direction that is likely to become ever more fruitful as increasingly sophisticated animal-like robots, particularly soft robots, begin to emerge.

### Do virtual robots count?

Various kinds of virtual entities, usually programs sat on a server in the midst of the internet, are often rather confusingly referred to as robots or "bots:" chat bots, virtual agents, AI-powered characters in computer games, even programs that just do some automatic data analysis. These are not robots. In this book the word robot will always refer to real physical devices you can point at and touch. Simulated entities with some robot-like properties, often rendered in computer graphics, I will refer to as virtual agents. However, there is often overlap in the technology underlying robots and virtual agents, and the line between the two can be fuzzy.

Techniques used in intelligent robotics have often been transferred directly into virtual applications. The ways in which some simulated characters move and act in computer-generated special effects in Hollywood blockbusters or in computer games draw on such methods. As in most areas of modern engineering, computer simulations are often extensively used during the development of physical robots and robot control systems. These are tremendously useful for trying out ideas and refining designs without the cost of building numerous prototypes. However, simulations are not reality, and something that works in a seemingly ultra-realistic simulation will not necessarily work in the always unforgiving, always noisy, always uncertain real world. This is not a problem for games and entertainment where there is no intention to transfer to the physical world. In that case it makes sense to cheat and bend the rules as much as we like as long as the result looks good to the player/viewer. But for a simulation used in the development of a real physical robot, the differences between the virtual and the real must be properly appreciated or the thing will never work as intended in the world.

So no, for the purposes of this book, virtual robots don't count, but they can be very useful in their own right and as stepping stones toward the real thing.

### What can robots do?

Currently, robots can do a surprisingly wide range of tasks, as long as they are relatively limited and tightly defined within narrow boundaries. Large, powerful industrial robots can act with great speed and precision as they spray paint over cars, or weld together components, or dispense adhesives, lubricants, or sealants with fabulous accuracy during fabrication processes, or select, move, and assemble parts on a production line, or inspect and measure components, or finally pack and load finished products, to mention just a few of the regular uses they are put to. It is this never-tiring power, speed, and precision that has enabled big increases in productivity and quality in many sectors of manufacturing. Behind the robot actions are sophisticated actuators and control programs that specify exactly what the robot must do. As we've learnt, even when such robots use complex sensing to help guide and refine their actions, they are doing so in a rigidly pre-defined way. They do not have anything that could be called initiative or intelligence. As well as being programmed for specific jobs, they are often physically designed for a single specialist task. A welding robot can't suddenly learn to be a paint sprayer.

More recently, companies such as Rethink Robotics have introduced a new breed of industrial robots that are intended to be smarter, more flexible, and easier to program for new tasks. Most importantly, they are designed to work collaboratively alongside humans rather than replace them, and hence are safe to be around. Traditional industrial robots are fast, heavy, and dangerous, and so are usually caged off to prevent accidents. Rethink "cobots" use screen-based animated cartoon faces to help communicate with their co-workers, and can be programmed without specialist knowledge. For instance, the robot can be taught what to do by simply guiding its multi-jointed arm to perform a specific job.[16]

Mobile autonomous robots can happily perform such tasks as floor cleaning (home vacuuming, industrial scouring and polishing), grass cutting, simple security, surveillance, and monitoring, and navigating from A to B while avoiding hazards. The latter is a very useful behavior for a variety of tasks including driverless vehicles, robot farming, moving objects around warehouses, and inspecting structures and buildings. Some of these robots use control systems as constrained and prescriptive as those for industrial manufacturing robots, and thus can only be used in well-defined environments in which there will be no surprises. Other are more adaptive, learning about their environment as they go. For instance, the more advanced robot vacuum cleaners build up maps of the home they are cleaning which help them to go about their business in a more efficient way, rather than just randomly blundering around until they have more or less covered the area. A number of large institutions are starting to use robots to give guided tours of their premises. As well as using internal maps to keep track of where they are, such robots typically convey pre-scripted nuggets of information to their guests as they take them on a pre-defined route. For instance, since 2018, some of the museums of the Smithsonian Institute in Washington, DC have had robot guides on duty.[17] As well as taking groups on tours, there have been experiments with using the robots to sonify paintings—that is, turn them into a synthesized soundscape—to enhance the experience of visually impaired visitors.

Various robots, usually humanoid or semi-humanoid, are starting to appear in the reception areas of some corporate headquarters, where they are used to greet people and respond to simple queries and process visitor information. Similar robots are used in some restaurants in Japan to take table bookings,[18] and are beginning to be utilized in simple teaching, caring, and companion roles. They generally act as mobile user interfaces, playing canned phrases through their speakers and interacting via touch-screen menus on a tablet fixed to their chests. Many are little more than attention-grabbing novelties, whose role could just as well be achieved with a wall-mounted touch screen; but research is ongoing on more advanced versions that attempt to recognize facial expressions, understand speech, and interact in much richer ways.

Fully or partially remote-controlled robots are used for dangerous tasks such as bomb disposal or the movement of nuclear material, or the inspection of hazardous areas, for instance after major accidents or serious chemical contamination. Specially designed medical robots, which are also usually remote controlled, are increasingly used for precision surgery, making certain kinds of procedures more efficient, and allowing the use of miniature instruments and smaller incisions, leading to less trauma, pain, and blood loss, thus promoting faster healing and less chance of infection. However, at the moment, the cost, and space and maintenance needs of such machines mean that they tend to make surgery more expensive, and they are not without their own risks.[19]

In the following chapters many other examples of what robots can do, and how they do it, will be explored.

### What can't robots do?

Plenty. In fact many of the behaviors you might imagine are possible, or almost possible, from the over-hyped claims that often appear in mainstream and social media and promo videos, are not, and won't be for some time. Present-day robots are already very impressive for dumb, repetitive, highly circumscribed work. For those kinds of tasks they far exceed humans in terms of strength, speed, accuracy, and repeatability. But the holy grail is general, intelligent, adaptive behavior. This would involve robots learning, through their own volition and without prompting, to generalize over multiple completely different problems, transferring skills and knowledge. That turns out to be much more difficult to achieve.

The blunt truth is that we are a long, long way from being able to instill our robots with anything like sophisticated general intelligence. The robot tasks mentioned above require either basic, highly specialized intelligence or no intelligence, or are human-controlled. This is because currently basic and narrow is the best we can do in terms of robotic intelligence.

Despite impressive advances in AI methods for handling language, robots cannot currently engage in proper, extended conversation. Some can answer specific kinds of questions by trying to look up the answers in search databases, or engage in largely pre-scripted exchanges, but they cannot take part in freewheeling chit-chat or witty banter, let alone initiate it. For the narrow tasks that they can learn—such as some kinds of visual pattern recognition—in general they cannot learn rapidly from just a handful of examples, as we can. (They usually require exposure to numerous training samples.) Robots do not have emotional empathy or imagination in any but the crudest sense. Their ability to improvise is as yet very limited. The average human has learnt to perform many different tasks—probably hundreds, even thousands—sometimes doing several at once. We can readily generalize and usually figure out what to do in completely novel, complex situations. We can quickly absorb new knowledge and learn new skills by building on past experience. Robots can do none of this. Robots cannot make sophisticated judgments of character or recognize imminent danger from an unexpected source never encountered before. Robots are not self-aware; they cannot realize that their allotted task is an outrageous affront to civilization and decide to do something completely different instead.

On top of all this, robots are reliant on their power source: they either need a lead connecting them to a fixed external power source such as mains electricity, or must rely on on-board batteries, sometimes with additional power-generating technology such as solar panels or even an internal combustion engine. For fixed robots, such as industrial arms, being tethered to a power source is not an issue, but mobile robots must carry their power around with them, and this can become a problem. The trade-offs between present-day battery size, power, and weight and overall robot size, weight, actuator power, and agility can quickly become very tricky, limiting the scope and duration of possible behaviors. For large, heavy robots, such as autonomous vehicles, it is less of a problem than for smaller robots where strength, speed, or durability is required.

On a more basic level, if a robot saw a broken, disassembled wooden stool with only two (unattached) legs, half covered by a dusty sheet, in the shadowy half-light of an attic it would not be able recognize it as a three-legged stool in need of repair. It would not be able to work out how to fix it with the random set of tools buried somewhere on the other side of the room. It certainly would not realize that it belongs to the same class of objects as a plushly upholstered sofa. A robot would have great difficulty in recognizing your face if you covered most of it with your hand. In contrast to children, a robot cannot learn a new kind of complex game after rapid explanation of the rules with a few words and lots of quick hand movements. A robot cannot be given vague, contradictory, almost entirely incorrect instructions and yet still be expected to successfully complete the required task (and yet we often manage this feat).

So, although some of our robots can rightly be said to be autonomous or intelligent, they are so only in a rather minimal, restricted way—which is not to say that they aren't useful, or that in time we won't learn how to make them more intelligent.

Robots may not yet be capable of anything even vaguely approaching the complexity and generality of human-level intelligence and adaptive behavior, but they are capable of very specific, limited versions of behaviors that approximate, and sometimes even surpass, aspects of lower forms of animal intelligence. Make no mistake: this is impressive. In well-controlled, benign conditions the best modern robot visual recognition systems can even approach human levels of performance for very specific tasks (such as spotting a cyclist), but once the light becomes murky, or heavy snow starts to fall, or the camera is shaken around a bit too much, the robot's performance plummets while the human's dips just a fraction. Our visual system has evolved to perform incredibly well over a huge range of conditions and to rapidly adapt to an almost limitless set of tasks. This kind of robustness and generality is a distant goal for robot perception.

To understand what is going on in robotics today, it is important to know something of the relevant history. The early history is surprising and very revealing—a major impetus has always been fiction: first the theatre and the novel and now Hollywood. A strange symbiotic relationship between robot fantasy and robot reality has existed from the start, and has helped to create the often uneasy images of popular imagination, as well as providing the raw inspiration for key technical developments. An appreciation of the origins of the core ideas behind today's machines will help to reveal their essence.

# 3

# SOME HISTORY

## Where did the idea for robots come from?

While stories of artificial human-like creatures go back at least to the myths of ancient Greece, the notion of embodied mechanical intelligence was, quite literally, thrust center stage in the years between the World Wars. In 1921, Karel Čapek's play *R.U.R.* (*Rossum's Universal Robots*) introduced the world to robots, in the process forging the associated myths and images that now permeate our culture. It was a worldwide smash hit, capturing the popular imagination as well as sparking much intellectual debate.[1] The play, with its roots in the dreams and folk tales of old Europe, blended with the nineteenth-century science fiction stories of Mary Shelley and H.G. Wells, told of the mass production of artificial humanoid workers on an isolated island. In a faint echo of the legend of the Prague Golem, these robots are created using some sort of biochemical process acting on a "living jelly" from which their parts are made. They are sold throughout the world as cheap labor with a limited lifespan. After a while they develop aggressive emotions and, realizing they are physically and mentally superior, the robots rise up and destroy the human race. The play ends on a more positive note when two robots develop feelings of love and respect toward life, and each other, becoming almost indistinguishable from the humans they have replaced. They venture out into the world as the new Adam and Eve. Elements of the plot will sound familiar because they have been reformed, recycled, and repurposed ever since, appearing again and again in literature and film. The play also sowed the seed for the scare stories about robots taking over the world which have appeared regularly in the popular media for at least the past six decades.

Karel Čapek had difficulty in deciding what to call the artificial workers until his brother, Joseph—a renowned Czech painter—coined the word robot. It is derived from the ancient Czech word *robota* which means repetitive drudge work. As the play was performed in more and more countries, the term robot quickly became absorbed into languages all over the globe.

Before Čapek's play—indeed going back as far as first-century BCE Greece[2]—many ingenious mechanical automata had been constructed, that is machines capable of following predetermined sequences of movements and operations. These included a chess-playing Turk and a flatulent duck in the 1700s. Such machines can perhaps be thought of as distant ancestors of today's robots, but—although some represented incredible feats of engineering for their time—they were mainly intended as theatrical demonstrations, akin to magic tricks, designed to delight, amaze, and above all impress.

In the eighteenth century, humanoid automata—also known as androids—were very popular: mechanical dolls dressed in elaborate costumes and designed to perform an impressive task to wow guests after dinner in some grand house, or to stun paying visitors to a public exhibition. Most were conceived as parlour novelties and wealthy aristocrats' toys, rather than serious tools. Androids that played a tune or two on a musical instrument were a guaranteed showstopper. In 1738 the virtuosic French artist-engineer Jacques de Vaucanson constructed a humanoid flute-playing automaton. It actually blew through its mouth and had a metal tongue to shape the notes in the way a human player does. Vaucanson made a detailed study of flute players in order to replicate fingering techniques and air flow in his android. The fingers were articulated by intricate systems of chains and levers hidden inside the arms and body. A flow of air was provided by several sets of bellows feeding into a pipe that led up to the mouth. The whole thing was driven—like all automata of the period—by the technology of clocks and music boxes. Fantastically complex assemblies of gears, cranks, and rotating discs with numerous precisely shaped and placed cams together controlled and coordinated the hidden internal mechanisms. (Cams are protuberances that would work sequences of levers or other mechanical linkages as they came into contact with them through the disc's rotation.) The flute player was the talk of Paris for months. It played extremely well, but it did have shortcomings that made it clear to professional musicians that this was no hoax—no human flautist would make the harsh sounds that occasionally came from the automaton;[3] the android really was producing the music, not a man hidden somewhere in the room.[4]

Vaucanson's most famous automaton was his mechanical duck. The exquisitely crafted bird could flap its wings in an uncannily life-like way (there were more than 400 moving parts in each wing), pick at its feathers while grooming itself, drink water, appear to digest its food, and defecate. The duck, complete with a rubber intestine, was supposed to accurately demonstrate digestion, but actually contained a hidden compartment of "digested food" which was pushed out as it defecated. The duck "ate" a mixture of water and seed and excreted a blend of bread crumbs and green dye that looked very like real duck excrement. In developing the duck's intestine, Vaucanson is believed to have invented the first ever flexible rubber tube. Not surprisingly, pushing the various mixtures through the duck's innards led to occasional bouts of flatulence.

These machines were products of the Age of Enlightenment—where scientific thought and method came to the fore—acting both as entertainments and provocative expositions, feeding into rationalist debates about the mechanistic nature of life, including mankind.[5] But no one worried about them rising up and taking over the world. They were viewed primarily as animated works of art.

One magnificent animal automaton from this period that has survived and can be viewed today, working in all its glory,[6] is Joseph Merlin's Silver Swan. It was made in collaboration with James Cox, and is one of the most beautiful machines ever constructed. The swan dates from 1773 and was originally the main crowd-pulling exhibit at James Cox's "Mechanical Museum" in London. The swan is life size and sits on a shimmering surface of "water" made from twisted glass rods through which exquisitely crafted little silver fish can be seen swimming and bobbing. The whole thing is fringed by a border of artfully drooping silver leaves lying at the water's edge. The machine is controlled by three interacting clockwork mechanisms that power its magical display. Once it is set running, the glass rods rotate, giving the impression that the swan is gliding over a glinting, moving surface. The neck then begins to articulate, moving the head gracefully from side to side, as if preening. The swan appears to spy a fish under the surface and, extending its neck fully, plunges its head down into the water. A moment later the head slowly comes back up with a fish in its beak. After a few motions of the beak and a toss of the head, the fish is swallowed down and the automaton returns to its starting position. All the while an integrated music box plays. When I first saw it in action, nearly 250 years after it was built, I was shocked by how deeply impressive it still is. Although not quite as complex as some other automata of the same period, there is something about the way it embodies a fusion of the clockmaker's, jeweler's, and automaton builder's arts that sets it apart: a transcendent blend of art and engineering.

Although these eighteenth-century automata were marvelous spectacles, it was apparent to most onlookers that they were slaves to their clockwork. Few would mistake them for sentient beings, not least because some automaton builders delighted in revealing the inner mechanisms of their creations.[7] They could perform one set of motions very well, over and over again. Some were designed with replaceable cam discs so could be "reprogrammed" (by a new combination of cams) to perform different tasks, but even then there was no sense in which they perceived the world around them and reacted to it in the manner of modern autonomous robots; they just blindly followed a script fashioned from cogs, springs, and rotating cam shafts. Then, in 1770, along came the chess-playing Turk (known as the Automaton Chess Player or Mechanical Turk),[8] which seemed to possess real intelligence.

Built by Wolfgang von Kempelen to entertain the mighty Empress Maria Theresa, ruler of the Habsburg dominions, which included much of central Europe, the automaton chess player consisted of a wooden cabinet on which sat the upper half of a life-size mechanical man, dressed in luxurious robes, as worn by some Ottoman Turks, and topped with a turban. In his left hand he held a long smoking pipe, and his right hand rested beside a chess board, complete with ivory chess pieces. The Turk's head would move as if surveying the state of the game, its facial expression would change to register intense concentration, and it could pick pieces up and move them around the board. Human challengers would sit on the opposite side of the board and try their best to beat the machine. They rarely did. For much of the next 68 years the chess-playing automaton was exhibited all over Europe and then in America, becoming very famous in the process. Only top-rated players were able to beat it, and even then not without some effort. The machine appeared to possess formidable cognitive and reasoning powers, at least in relation to chess. It could even tell if an opponent was trying to cheat with illegal moves, responding by moving the piece back to its original position. It seemed to have excellent visual perception and hand–eye coordination: good enough to quickly assess the state of the board and deftly move the pieces. Many suspected it must be a fraud, but before each exhibition the master of ceremonies (originally Kempelen and then later others as the machine changed ownership) made a great show of opening the sides of the cabinet, revealing various internal mechanical mechanisms and apparently showing a clear path all the way to the other side, convincing viewers that there was no possibility of a hidden operator squatting inside the machine.

But of course it *was* all an elaborate hoax. A triumph of cabinet making, trickery, and misdirection: the art of the magician. An operator *was* indeed squashed inside but on a sliding seat such that as the various doors and drawers were opened he could always stay out of sight. Much of the internal mechanism revealed by the master of ceremonies had no actual purpose. There were strong magnets hidden in the base of the chess pieces which attracted magnets on the underside, allowing the operator to keep track of his opponent's moves. The operator, who was a highly proficient player, operated a peg-board attached to the Turk's arm via a system of mechanical linkages (similar to a pantograph) which allowed him to make precise movements of the arm around the board; a lever enabled him to open and close the hand as he moved the chess pieces. There were also various hidden ways for the operator (who worked by candle light) and master of ceremonies to communicate. There was plenty of hugely impressive perception, reasoning, and motor coordination, but it was all human.

It was not until 1857 that the Mechanical Turk's secrets were finally fully exposed. The machine's final owner was John Mitchell, a wealthy physician, who donated it to a museum in Philadelphia. In 1854 a fire swept through the museum and destroyed the fake automaton. Mitchell's son, Silas, also a physician, decided it was now time to reveal all. In 1857 he published a series of articles for *The Chess Monthly* that explained the ingenious illusion.[9]

## When were actual robots first built?

From the late 1920s through to the 1940s various machines started to appear that were described by their inventors as robots (after Čapek). They were mainly "tin men" who performed a trick or two. Two of the most famous were Eric the Robot and Elektro the Moto-Man (see Figure 3.1).

Eric was built in 1928 by WH Richards, a retired British journalist, businessman, and self-taught engineer, in collaboration with Alan Reffell, a skilled mechanic.[10] Richards was Acting Secretary of the Society of Model Engineers whose annual exhibition was to be opened in London by the Duke of York. When the Duke became unavailable, Richards decided to build a mechanical man to launch the exhibition instead. Eric was life-size, made of gleaming aluminum, and, as many commented at the time, looked like a knight in shining armor. As a nod to Čapek's play, he had the letters RUR emblazoned on his breast-plate. Responding to voice commands, Eric could stand up,

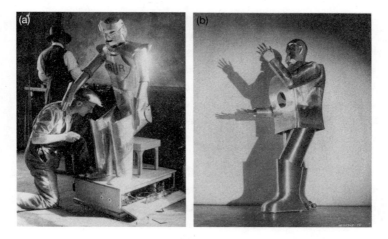

**Figure 3.1**  Left: Alan Reffell working on Eric the Robot, with WH Richards at the bench behind. Right: Elektro the Moto-Man at the 1939 World Fair.

*Source:* Left—Edward G. Malindine/Getty Images; right—Manuscripts and Archives Division, The New York Public Library.

sit down, raise his arms, bow, and move his head from side to side. His eyes shone red and a high-voltage source created fearsome blue sparks between his jagged teeth. His feet (and lower legs) were firmly fixed to a wooden box in which was an electric motor. Further motors, electromagnets, pulleys, and cables inside his body enabled his various pre-programmed movements to be triggered and switched between.

By modern standards, and indeed the standards of the best eighteenth-century automata, his movements were rather clunky.[11] But he did have one trick that brought him firmly into the twentieth century: he could speak to his audience. Apart from delivering a speech to open the Model Engineers' exhibition, he answered questions from the audience, moving his arms and head as he did so. He performed for the whole week of the exhibition, rising and bowing and conversing thousands of times. News of the machine quickly spread, bringing huge crowds flocking to witness him for themselves; numerous articles on his exploits appeared in the press, some exaggerated and inaccurate—for instance claiming that the robot walked across the stage to a lectern to give his speech. Like Vaucanson before him, Richards was happy to reveal much of the inner electro-mechanical workings, but he kept the details of how Eric spoke a closely guarded secret. Richards did tell interviewers that Eric could not see or think and had a fixed menu of 50 or 60 possible answers to questions, including "I do not know, sir" (or madam, depending on who asked the question). The fact that the gender of the questioner could be correctly identified is a clear indication of an off-stage accomplice. Richards also claimed that he was working under license from Marconi's Wireless Telegraph Company (pioneers of radio communication and broadcasting) and was using "the most advanced methods of radio control." The voice commands may have automatically triggered some kind of remote control, or more likely this was coordinated by someone off-stage. It is possible that the "man's voice" heard through the robot's "voice box" was produced by radio communication, allowing an assistant with a microphone to speak through Eric; or there might have simply been a hidden cable from an amplified microphone connected directly to the voice box, which was presumably some kind of loudspeaker. It is also possible that the answers were all pre-recorded and somehow triggered by the assistant behind the scenes (although Richards stated that "[Eric's] speech is produced neither by phonograph record or talking film"[12]).

Eric seems to have been a highly successful mix of engineering, showmanship, and illusion. He went on to become world famous, touring the globe as "Eric the amazing robot: the man without a soul." Richards exploited the robot's potential for entertainment and marketing but also insisted that some of Eric's engineering principle and mechanism could be further developed to create machines capable of performing useful manual work involving specific repeated movements. An improved version, named George, was developed a few years later and toured until 1936.

Elektro the Moto-Man's origins were even more solidly rooted in marketing. In the late 1920s Roy J. Wensley, of Westinghouse Electric Company's Pittsburgh plant, invented a clever control device, known as the Televox, which allowed engineers to remotely change the setting at an electrical substation over a telephone line.[13] A Televox unit at the central power station could send a variety of tones over a telephone line to another unit at the substation which could interpret them and initiate the required actions, such as resetting various switches. The Televox units were based around networks of electrical relays (electromechanical switches that can be automatically opened or shut by the application of current). Wensley could see many applications for his device and asked Westinghouse to build a publicity campaign for the Televox. They turned him down. Undeterred, he used what small budget he had to build a portable demonstration unit and start his own advertising. He guessed the raw box of relays and wires would be unlikely to appeal to the general public, but he knew what would: a robot.

Tapping into the growing popularity of science fiction and taking advantage of the fact that so many of us were, and still are, suckers for robots, Wensley quickly created Herbert Televox. Herbert was a crude 2-D humanoid figure made from board, with arms articulated at the shoulder and elbow. A cut-out in his torso revealed the inside of a Televox unit. The controller could be used to trigger movements in the arm, thus enabling Herbert's party piece: unveiling a portrait of George Washington. Herbert immediately caught the imagination of reporters and the wider public. While Eric was filling British newspapers and magazines with exaggerated stories of his prowess, exactly the same thing was happening in the USA with Herbert. This helped spark an interest in the possible applications of such technology to the automatic control of military hardware and other machinery. Although this did not immediately come to fruition, it sowed the seed for later developments. Westinghouse could now see the huge publicity potential and allowed Wensley to continue with other robot projects, working with an expanding team of engineers. Mr. Telelux, Katrina Van Televox, Rastus Robot & Willie Jr., and Willie Vocalite followed over the next few years, each showcasing various Westinghouse technologies. In 1938 came Elektro the Moto-Man, the most successful of the lot.

Elektro was the star of the 1939 World Fair in New York. Standing 2.1 m tall, he was built from aluminum. Elektro could do a variety of tricks, including counting on his fingers, singing, talking, walking, and smoking a cigarette. The imposing, barrel-chested robot was stuffed full of relays, motors, and steel gears, as well as electrical sensors (see Figure 3.1). Actually, to say the robot walked is a bit of a stretch—it shuffled along on a set of wheels by slightly bending its left knee and dragging its right leg behind.[14] Its speech was all pre-recorded and played from a set of records. Designed primarily as a crowd-pleasing novelty act, a carefully rehearsed presenter would interact with the robot by following a rigid script. However, some of Elektro's technology was cutting edge and paved the way for later innovations. The robot used a—for the time—very advanced voice command system. The presenter would speak commands clearly and slowly into a telephone, which was connected to an off-stage machine that was able to measure the timings of the syllables which followed set patterns, acting as a code. The code triggered a flashing light signal which was sent across the room to a photoelectric sensor in the robot's chest. The sensor converted the light signal into electric pulses which were processed by the relays and other circuitry to trigger the required movements of the robot's internal mechanisms.

Although they were built from shiny metal and powered by electromechanics rather than clockwork, Eric and Elektro were in many ways updated versions of the automata that had entertained the wealthy two centuries earlier with their collections of pre-scripted behaviors. Marketing, trickery, and showmanship were very much to the fore, but unlike the eighteenth-century automata, these machines were genuine robots, according to my definition given at the start of Chapter 2. They were obviously able to perform actions in the world (as were the ancient automata), but, unlike the automata, they made use of senses. Both responded to voice commands: detecting these commands involved some level of sensing; their subsequent behavior can therefore be regarded as a limited form of acting in reaction to external signals. In Elektro this sensing ability was quite sophisticated, as outlined above; in Eric it was more crude and seems to have involved responding to particular frequencies.

Perhaps most importantly, both caught the imaginations of future generations of engineers and opened their eyes to the possibilities of robotics. They also helped to further cement the popular image of a robot as a hulking metal man.

### When did genuinely autonomous robots first appear?

Entertaining as they were, the robots of the 1920s and 1930s were far from autonomous. They followed precisely orchestrated behaviors, unable to deviate from their built-in actions. It would be another decade before robotics moved off in another altogether more interesting direction.

   Readers may be surprised to learn that the first proper autonomous robots were built not in a commercial or university engineering laboratory, but in the home workshop of the swashbuckling British neurophysiologist Grey Walter, who wanted to use them to learn more about how the brain works. The story forms a very nice example of how open-minded interdisciplinary science can take unexpected paths, and profoundly influence later generations.

   William Grey Walter (always known as Grey) started his research career in 1931 at Cambridge University, working on advanced problems in neurophysiology under the supervision of Bryan Matthews and Nobel laureate Edgar Adrian, two giants of the field.[15] (Neurophysiology focuses on the workings and organization of the nervous system, including the brain, and is now part of the wider modern field of neuroscience.) Grey—who had started building radio sets when he was nine years old—soon became a master of the advanced experimental techniques of electroneurophysiology, which included using cutting-edge electronic amplification to detect neural signals. It was this technical prowess that pitched Grey into his next role when he left Cambridge in 1935, and which would eventually lead him to leave his mark on the history of robotics.

   Grey, who seemed to be perpetually followed by a vague cloud of scandal,[16] went to work as a neurophysiologist at the Maudsley Hospital, London, under the renowned neurologist Frederick Golla. He was employed to investigate possible clinical applications of the *electroencephalogram* (EEG)—noninvasive recordings of electrical activity in the brain made through electrodes placed on the scalp. First demonstrated in the 1920s by Hans Berger, the EEG is a series of wavy lines tracing the voltages between the electrodes. The EEG generally shows characteristic, regular oscillations: brain waves.

Building his own EEG recording machines, including the first ever portable device, Grey made many important discoveries, including showing that characteristic EEG patterns could be used to indicate and locate brain tumors, and then later that there was an EEG signature for epilepsy which could be used in the diagnosis and study of the condition.

The outbreak of the Second World War sparked a rapid spread of clinical applications of EEG throughout the UK, particularly in the diagnosis of head injuries. Grey did much to spread knowledge, practice, and equipment throughout the country. In order to provide more information to aid diagnosis, in spring 1943 he succeeded in developing a machine that automatically extracted the main frequencies present in an EEG, the first of its kind. His analyzer worked continuously and in real time.[17] Later versions of the machine were used in hospitals and research labs all over the world.

It was Grey Walter's frequency analyzer that led to a momentous wartime meeting, sometime during 1943–44, that sparked the idea for one of the most important landmarks in robotics. He was contacted by Kenneth Craik, a brilliant and driven psychologist at Cambridge University, who had been seconded onto many secret wartime research projects, mainly related to improving the way people interacted with military hardware—from radar screens to artillery control mechanisms. Craik had some complex data on air gunners' aiming errors and wondered if Grey's analyzer could help him understand it. Craik visited Grey (by then based at the Burden Neurological Institute, Bristol) to study the analyzer in detail.[18] The two scientists spent hours in wide-ranging discussions about the brain, the generation of behavior, and advances in technology. They had much in common, including a deep love of the design and construction of scientific apparatus.

Craik's fascination with mechanical devices, his considerable engineering abilities, and his work with very early specialized electronic computers[19] no doubt informed the radical thesis of a book he had just published, *The Nature of*

*Explanation.*[20] This work was to have a galvanizing effect on the later development of cybernetics—the scientific movement that foreshadowed AI. He proposed that the nervous system is a kind of prediction device—that neural mechanisms, somehow acting as "small-scale models" of external reality, could be used to try out various alternatives, reacting to future situations before they arise, and this was at the center of their enormous adaptive power.

Craik suggested that such predictive power is not unique to minds. Indeed, although the flexibility and versatility of human thought is unparalleled, he saw no reason why, at least in principle, such essential properties as pattern recognition and memory could not be emulated by a man-made device. He claimed that the best way to think of the human mind is as a kind of machine—a machine that is geared up to anticipate events.

He proposed that the proper study of mind is an investigation of mechanisms capable of generating intelligent behavior *both* in biological *and* non-biological machines. The mechanisms of natural intelligence can only be said to be understood when it is possible to specify them in enough detail to build a working machine. He advocated the "synthetic method," where physical devices were built to explore biological mechanisms. Because it is extremely difficult to systematically study biological processes (such as the neural mechanisms active while an animal or human is engaged in some behavior), he proposed building artificial systems, based in some way on the biological system, but greatly simplified. Such systems *could* be systematically studied, as all components are readily accessible and controllable by the experimenter.

Craik and Walter's conversation soon came round to scanning. Walter had a scanning hypothesis relating to the function of a particular type of brain wave—the alpha wave. He likened it to the scanning radar systems that had become visible in many strategic locations. Craik was very familiar with scanning mechanisms from his involvement in radar and automatic gun control research. The latter was an example of goal-seeking behavior, especially when it involved following a moving target. Grey later described the thrust of their discussions:

> Goal-seeking missiles were literally much in the air in those days, so in our minds, were scanning mechanisms.[21]

The pair cooked up a scheme where these two processes, scanning and goal-seeking, were the core elements of a "working model that would behave like a very simple animal." They talked about the idea of constructing "a free goal-seeking mechanism"—what we would now call an autonomous robot—which would use sensors that scanned the environment in order to locate a goal to which it would then move. They excitedly discussed how such an artificial creature could be used to investigate the essential properties of neural mechanisms that might underlie such behaviors. This could be done by building an artificial—presumably electronic—nervous system, or series of nervous systems, which could be systematically studied. In other words, it was an ideal testbed for Craik's synthetic method.

Craik and Walter planned to collaborate on building such an artificial creature once the war was over. Tragically, Craik was killed on the last day of the war in Europe when he was knocked off his bicycle in a traffic accident. But the discussion stayed with Grey, and in 1948 the artificial animal became a reality when he completed the first of a series of machines he called "tortoises", machines that were destined to become a media sensation all over the world. The tortoises were the first ever mobile robots that were genuinely autonomous. The devices were three-wheeled and sported a protective "shell" (see Figure 3.2). They had a scanning light sensor, touch sensor, propulsion motor, steering motor, and electronic valve (vacuum tube)-based analog "nervous system."

**Figure 3.2** Grey Walter with Elmer and Elsie, two of his tortoises. A long exposure with candles on top of the robot bodies reveals the paths they traced as they moved around the room, attracted to each other and the glowing fire in the hearth.

*Source:* Larry Burrows/The LIFE Picture Collection via Getty Images.

### What did Grey Walter's tortoises do?

Grey Walter's intention was to show that even in a very simple nervous system (the tortoises had just two artificial neurons) complexity could arise if the possible interactions between its units were sufficiently rich. By studying whole embodied sensorimotor systems—physical robots that acted in the real world—in order to explore what kinds of mechanisms could underlie behavior in animals and machines, he was pioneering a style of research that was to become very prominent in AI and related fields many years later, and remains so today. From Easter 1948 onward,[22] Grey and his second wife, Vivian Dovey, a radiographer, built the first tortoises, Elmer and Elsie. Elmer was finished toward the end of 1948 and Elsie a few months later. In 1951, his technician, Mr. W.J. "Bunny" Warren, built six new tortoises to a much higher standard so they could operate reliably over longer periods without needing running repairs and recalibrations. Three of these tortoises were exhibited at the Festival of Britain in 1951,[23] where they were a huge hit; others were regularly demonstrated in public throughout the 1950s. The robots were capable of phototaxis (steering toward a light source), which enabled them to find their way to a recharging station (their "hutch") when they ran low on battery power. He referred to the devices using the mock zoological name *Machina speculatrix*, after their apparent tendency to speculatively explore their environment.[24]

Walter was able to demonstrate a variety of interesting behaviors as the robots interacted with their environment and each other. In one experiment he placed a light on the "nose" of a tortoise and watched as the robot observed itself in a mirror. "It began flickering," he wrote, "twittering, and jigging like a clumsy Narcissus." Walter argued that if this behavior was observed in an animal it "might be accepted as evidence of some degree of self-awareness."[25] Later tortoises, with more complex nervous systems, were able to learn to perform particular behaviors depending on the sound picked up by a microphone on their body. (Grey used a two-note policeman's whistle to train them: one note became associated with one behavior, the other with a different behavior.)

Walter's robots became globally famous, featuring in newsreels, television broadcasts, and numerous newspaper and magazine articles, capturing the imagination of the public and fellow scientists alike. They have been acknowledged as a major early influence by a number of leading robotics researchers of later generations, including Rodney Brooks, formerly a robotics professor at Massachusetts Institute of Technology (MIT), and director of the AI Lab there, and more recently a fabulously successful robotics entrepreneur (he was founder and original CTO of iRobot, the company behind the multi-million-selling Roomba robot vacuum cleaners). Brooks built his first robot—a version of *Machina speculatrix* using transistors rather than valves[26]—after coming across Walter's 1953 book *The Living Brain*.

### How did the tortoises work?

The tortoise chassis was in the form of a tricycle. The two rear wheels provided stability, and the single front wheel was controlled by two motors: one to drive the wheel and one to power the steering. In *Machina speculatrix* there were two sensors: the light sensor (a photoelectric cell), which sat above the front wheel steering mechanism and rotated with it, and a touch sensor, which was activated when the tortoise's shell knocked against an obstacle. The light sensor can be seen in Figure 3.2; it is housed in the periscope-like tube sticking up on the top of the robot. Sensory inputs and motor outputs were connected via an electronic nervous system of two artificial neurons built from relays, electronic valves (tubes),[27] and a few other components. The ways in which the neurons interacted with each other, and thus affected motor outputs, depended on the sensory input. There were four basic behaviors which interacted with each other.

The first behavior occurred when the light level picked up by the photocell was low. In this case, the front wheel was driven at low speed and the steering motor was driven at quite high speed. Because the steering motor was also directly connected to the light sensor, this made the sensor continuously rotate, scanning the environment. The overall effect in this state was for the tortoise to move in spiraling "explorative" patterns (see Figure 3.2 for behavior traces showing the path of the tortoise).

The second behavior came into play when the photocell detected a moderate light level. When that happened, the driving speed switched to high and the steering motor stopped altogether. This meant the light sensor also stopped scanning. Thus the tortoise moved quickly in the direction the front wheel happened to be set at when the steering motor stopped. If this was straight and in line with the light source, the artificial creature would move directly to the light. However, if the steering was at some angle, as was usually the case, the tortoise would move in a curve. The curved path would inevitably mean that after a while the photocell would be pointing away from the light source and the sensor reading would drop. Once that happened the original explorative scanning behavior would resume until the light was once more "found."

If the light level became high, for instance if the tortoise got close to a bright light source, the third behavior took over. In this case, the wheel was driven at high speed and the steering motor switched to low speed. This would make the tortoise veer off at high speed, taking it away from the light. The photocell reading would drop, and the first explorative behavior would resume.

If at any point the touch sensor was activated, because the tortoise had bumped into something, the fourth behavior would switch in. In this case, some clever circuit design meant that the two artificial neurons would now become interconnected in such a way that they created electrical oscillations. The oscillating signal thus created was used to alternately drive the steering motor and then the drive motor for short bursts. The result was that the tortoise would jiggle about, pushing and turning, until it was free of the obstacle.

The ingeniously designed nervous system allowed various combinations of these basic behaviors to emerge in interesting overall sequences.[28] You can clearly see the behaviors in the long-exposure photo shown in Figure 3.2. The overall "goal" was to seek out moderate light sources.

The crucial point is that there was no rigid script being followed, as in all previous robots. Rather, the sensors, motors, artificial nervous system, environment, and history of movements combined—sometimes in quite complex ways—to generate the observed behavior. Grey Walter had very carefully designed the nervous system to produce the desired kinds of outputs, but even he could not say in advance exactly what would happen. The tortoises behaved slightly differently every time—rather like real animals; changes in the environment could result in quite different behavioral patterns emerging over time. The tortoises were autonomous.

This unpredictability was one of the things that made the machines seem much more life-like than the clunky tin men of the preceding decades. Their resemblance to nature partly drove their immense popularity; people were fascinated by them, just as they are by animals. Although the media at the time often referred to them as robots, Walter hardly ever did. He used the term "artificial creature" or "artificial animal," partly to emphasize his scientific aims and partly to deliberately avoid comparison with what he saw as the commercial gimmickry and trickery of earlier robots, such as Elektro. Grey was quite a showman himself and was certainly not averse to publicity, but his scientific purpose was serious and he did not want to obscure it.

The tortoises were not the first phototropic (light-attracted) mobile robots. For instance, in 1928 Henri Piraux built a robot dog, Philidog, to promote photocells developed by Philips. The robot was demonstrated in 1929 at the International Radio Exhibition in Paris. Philidog's electronics sat on a simple frame with two drive motors attached directly through amplifiers to the photocells. A wooden dog-like body covered the mechanism, with the photocells fitting inside the head, acting as the eyes. When a light was shone at the right eye the robot would move in that direction, with the same for the left eye, and when both cells were illuminated the robot would move directly forward. But there were no autonomous, self-generated behaviors. Instead, the demonstrator moved the robot about by carefully shining a torch at the photocells—a form of remote control.[29] A number of other mobile robots appeared during the 1930s and 1940s, most operating in a similar fashion; some were constrained to move on toy railway tracks and were reliant on externally controlled cues. The tortoises were the first genuinely autonomous mobile robots.

In the 1990s Owen Holland, then of the University of West of England, led a team who restored a surviving tortoise and built two replicas, thus making it possible for new generations to experience the impressive capabilities of Walter's robots first hand.[30]

At exactly the same time as Walter and Dovey were building their tortoises, a new scientific movement was starting to emerge: cybernetics. The tortoises are now seen as the most famous of all cybernetic machines and, although always something of a maverick, Walter was to become one of the leading figures of this new field.

### What is cybernetics?

In 1948 the American mathematician Norbert Wiener published a book, *Cybernetics, or Control and Communication in the Animal and the Machine*,[31] whose title simultaneously named and summed up the ideas at the core of the emerging scientific movement he was spearheading. Cybernetics blended ideas from engineering, mathematics, and the brain sciences to try to find general, unifying ways of understanding how behavior is generated and information processed in brains, and how the same might occur in machines. In particular, how is intelligent, adaptive behavior produced in animals and how might it be developed in machines? How are the interacting processes in such necessarily complex systems regulated? How are they controlled to produce the desired outcome?

The ideas gathered together under the cybernetics banner had been developing for some time, with various intellectual strands gathering pace in the UK, USA, and elsewhere in the 1920s and 1930s. But it was the Second World War that proved the catalyst in bringing the different themes together and forging this new scientific movement. Especially in the UK, biologists and psychologists found themselves put to work on projects related to communications and radar. This had the effect of throwing them into research involving electronics and communication theory, where they worked directly with physicists, engineers, and mathematicians. This mixing of people and disciplines led to an important two-way flow of ideas that was later to prove highly significant in advancing the formal understanding of the nervous system, as well as in developments in machine intelligence and robotics.

Ideas on learning, self-regulation, and control in animals and machines developed in parallel, and largely independently, in the UK and USA throughout the 1940s, although Craik's ideas and Alan Turing's seminal work on the foundations of digital computing had become influential on both sides of the Atlantic by the mid-1940s. From 1945 onward interactions between the two groups grew. The American movement beyond Wiener's immediate circle included mathematicians John von Neumann, Claude Shannon, and Walter Pitts, along with neurophysiologist Warren McCulloch. In the UK a meeting on cybernetics organized by Grey Walter in summer 1949 helped spark the formation of the Ratio Club, an important interdisciplinary dining club for discussing cybernetic ideas. Organized by neurologist John Bates, the club included Walter and Turing as members, along with neurophysiologists Horace Barlow and William Rushton, psychiatrist Ross Ashby, and astrophysicist Thomas Gold.[32]

Cybernetics, then, was the forerunner of AI and cognitive science. For a decade or so it was one of the hottest topics in science. With its emphasis on biological influence and understanding, it focused on adaptive, self-organizing systems. Thus the roots of machine learning, neural networks, and much of modern AI, so important in contemporary robotics, lie in cybernetics. Through Walter's tortoises and later machines such as Wiener's moth[33] (similar to the tortoises but more limited), it also gave birth to autonomous robotics. In relation to machine intelligence, the field was largely replaced by AI from the 1960s onward. But its ideas did not die: key cybernetic principles relating to flow of information, control and self-regulation, and dealing with uncertainty strongly influenced important developments in engineering, biology, and economics that have persisted. As we shall see, since the 1990s, cybernetic ideas have made a strong comeback in AI and robotics.

## When did industrial robots start to be used in factories?

In contrast to Walter's biologically focused and rather exploratory research, the main thrust of robotics began to turn to more immediately practical applications during the 1950s. A major emphasis was on the development of robot arms, particularly for use on production lines. Rather gradually, over a period of about 20 years, robot arms and manipulators became more and more widespread in heavy industry.

Control of these arms would build on other strands of cybernetics that had also developed from military research during the Second World War. Work during the war on methods to accurately, automatically, and smoothly move heavy artillery and radar dishes, and to coordinate the movement of both in target-following systems, led to advances in feedback control (Wiener was involved in such work in the USA, as were several future members of the Ratio Club in the UK). The idea of feedback, in which part of a system's output, or effect on the environment, is fed back to it to enable self-regulation by making small adjustments proportional to the perceived error, is something we are all intuitively familiar with. If a ball goes further than intended, we throw it more gently next time; if it goes too far to the right, we aim more to the left. During the war a more formal understanding of feedback was developed. This theory was to become of central importance in cybernetics, both as a way of understanding self-regulating biological systems and for designing more effective control mechanisms for machines. The latter strand developed into the field of control theory. Better servomechanism—motor systems with error-correcting feedback to ensure accuracy—resulted and fed into the development of heavy-duty industrial robot arms. In 1959 a robot arm prototype developed by George Devol and Joseph Engelberger, known as the Unimate, was installed on an assembly line at a General Motors diecasting plant in New Jersey. By 1961, the Unimate 1900 series became the first mass-produced industrial robot arm. Very soon there were hundreds of Unimate arms at work in diecasting around the USA. Gradually the technology spread to Europe and then Japan, and to other areas of industry. Now there are millions of robot arms in use all over the globe.

The central goal of classical industrial robotics is to move the end of an arm (which houses an actuator such as a gripper, known as the end effector) to a predetermined point in space. This is generally approached by finding the required sequence of rotational forces, applied through the motors controlling the arm joints, such that the resulting arm configuration puts the end effector in the desired position.

### When did work on intelligent autonomous robotics start?

The focus of industrial robotics, dominated by arms, was on reliable, powerful machines that could be programmed to perform relatively simple repetitive tasks with incredible accuracy and speed without ever flagging. More interesting and challenging behaviors would require some sort of intelligence.

From the early 1960s the newly established field of AI began to take over from cybernetics as the dominant force in attempts to make machines smart. Although most of its roots were firmly in cybernetics, AI became less biologically inspired, less concerned with learning, adaptation, and self-organization, and more focused on approaches to "program in" intelligence, often based on systems of logical rules. From 1966 to 1972 the Artificial Intelligence Center at the Stanford Research Institute (SRI) conducted pioneering research on a mobile robot nicknamed "Shakey." The robot had a vision system which gave it the ability to perceive and model its environment in a limited way. Shakey could perform tasks that required planning, route-finding, and the rearrangement of simple objects. Shakey became a paradigm case for early AI robotics. The robot, shown in Figure 3.3, accepted goals from the

**Figure 3.3** Shakey the robot in 1970.

*Source:* SRI International.

user, planned how to achieve them, and then executed those plans.[34]

Shakey operated according to a theory of intelligence in which an internal model of the external world has to be built, maintained, and constantly referred to in order to decide what to do next. This idea in part stemmed from Craik's 1943 notion that the nervous system somehow creates "small-scale models" of reality in order to predict what action will be best in the current circumstances, although Craik was thinking much more in terms of dynamic neural systems rather than logic. Hence Craik inspired both the tortoises and the very different approach of early AI robotics. Shakey's world model was built from formal logic and was operated on by various different computational modules which underpinned perception, planning, and action in the robot.

### How well did these robots perform?

Shakey employed a sequential processing approach: a strict pipeline of operations. First, perceptual routines were used to build up and modify the world model based on sensory information, particularly from the robot's vision system. Then, reasoning routines created a plan of what to do next. Finally, the current step of the plan was converted into motor commands to control the robot. This process, in which the same sequence of operations—sense–think–act—was repeated over and over again, could be computationally very expensive, even in the carefully constructed environments in which the robot operated (these mainly consisted of large colored blocks of various regular shapes and sizes). Nevertheless, the general approach was very influential and it dominated for the next 20 years.

However, even though Shakey and robots like it were controlled by computers the size of a room (connected to the robots via radio links), the demands of the endless sequential sense–think–act loop were such that they could not operate in real time. They would often take tens of minutes or even hours to complete a single task such as navigating across a room avoiding obstacles, often pausing for minutes at a time between painfully slow movements as they tried to reason about what the next action should be, sorting through the various possibilities. By the mid-1980s a number of leading researchers from the main AI robotics centers were becoming more and more disillusioned with the approach.

### What happened next?

Hans Moravec, an influential roboticist who had done important work on the Stanford Cart, a project similar in spirit and approach to SRI's Shakey and which ran at about the same time, summed up such feelings:

> For the last fifteen years I have worked with and around mobile robots controlled by large programs that take in a certain amount of data and mull it over for seconds, minutes or even hours. Although their accomplishments are sometimes impressive, they are brittle—if any of many key steps do not work as planned, the entire process is likely to fail beyond recovery.[35]

Moravec went on to point out how this is in strange contrast to the pioneering work of Grey Walter and the projects that his tortoises inspired. Such early robots' simple sensors were connected to their motors via fairly modest circuits, and yet they were able to behave very competently and managed to extricate themselves from very difficult and confusing situations without wasting inordinate amounts of time "thinking." In conclusion, Moravec advocated making the most effective use of whatever technology is available in the present in order to be able to gradually build up experimental discoveries, rather than developing more and more complex reasoning systems that cannot be used in any meaningful way in real time in the real world.

The "classical" AI approach—reasoning with the aid of internal world models—was faltering in many areas of AI, not just robotics. Disillusionment with this state of affairs found a particularly effective voice in Rodney Brooks, who was developing an alternative vision of not only intelligent robotics, but also the general AI problem.[36] Influenced by Moravec, as well as by the unconventional work of Marc Raibert (who produced a wonderful series of walking, running, and hopping robots[37] and went on to found the robotics company Boston Dynamics), Brooks, along with his team at MIT, became central to a growing band of dissidents who launched a salvo of attacks on the AI mainstream.

In a move that conjured up the spirit of cybernetics, the dissidents rejected the assumptions of the establishment, instead regarding the major part of natural intelligence to be closely bound up with the generation of adaptive behavior in the harsh, unforgiving environments most animals inhabit. In a direct echo of Craik and Walter the investigation of complete autonomous sensorimotor systems—"artificial creatures"—was seen as the most fruitful way forward, rather than the development of disembodied algorithms for abstract problem solving, which had become the focus of most of AI by then. The central nervous system was viewed as a fantastically sophisticated control system, not a chess-playing computer. Hence it was claimed that the development of mobile autonomous robots should be absolutely central to AI. Vested interests were threatened, emotions ran high, and insults were traded. It was an exciting time!

At the heart of Brooks' and similar approaches were biologically inspired control architectures involving the coordination of several loosely coupled, relatively simple behavior-generating sub-systems, all acting simultaneously *in parallel*. Each had access to sensors and actuators and could act as a standalone control system (one system might be taking care of not bumping into things, one might be seeking out a target, one might be making sure the robot doesn't go into danger areas, and so on). The sub-behaviors were designed to interact in such a way as to produce a coherent overall behavior suited to the task in hand (very reminiscent of Walter's tortoises). Such an approach was shown to robustly handle multiple, sometimes conflicting, goals and could be expanded in a natural way—areas in which traditional methods struggled. With the unwieldly, strictly sequential sense–think–act pipeline discarded, the new approaches worked in real time in the real world.

Whereas traditional AI robotics had taken a top-down approach, starting from the assumption that relatively complex sensors and general-purpose reasoning would be needed for anything worthwhile, the new "behavior-based" robotics took a bottom-up approach, in which behavior emerged from interacting sub-systems reliant on relatively simple sensors and processing. Such work triggered the formation of the so-called New AI movement in the 1990s, still strong today. With its focus on the development of whole artificial creatures as an important way to deepen our understanding of natural intelligence and provide new directions for the engineering of intelligent machines, it pushed robotics back to the forefront of AI.

Adaptation and learning were major concerns of cybernetics because it was felt that a good way for machines to deal with complex and uncertain environments was to learn and adapt through experience (just as animals, including us, do). By largely ignoring this issue, traditional AI robotics (and AI in general at the time) had a big weakness: it was "brittle"—if anything unforeseen happened, such systems usually failed. In most cases the AI programmer attempted to anticipate all possible future interactions of their robot and make sure appropriate responses were "programmed-in." This could be made to work in very constrained, well-designed environments. But for a mobile autonomous robot operating in a noisy, changing, unpredictable environment, with fluctuating lighting levels and people wandering around, it just wasn't feasible—it was not possible to second-guess all possible outcomes in the face of noise and uncertainty. Consequently, when the robot encountered situations that didn't fit in with the assumptions implicit in its world model and reasoning algorithms, its inflexible logic-bound methods could not compute how to respond. Its logic would attempt to modify its world model, repeatedly examining and manipulating and remodifying as it flayed around desperately trying to resolve the conflicts in its reasoning. Eventually it ground to a halt, defeated. Unburdened with complex reasoning, behavior-based robots—like the tortoises before them—just got on with the job as best they could.

As mentioned earlier, some of Walter's second-generation tortoises from the early 1950s learned to respond to stimuli, new behavior being "laid down" by changes to the electronics of the nervous system. This approach was revived; increasingly sophisticated techniques for learning and adaptation in robots began to be developed in the 1990s and 2000s. AI and intelligent robotics had once more moved much closer to their cybernetic roots.

The explosion in biologically inspired robotics that Brooks' work helped to fuel brought forth numerous interesting strands of research, many of which are still very active, and pushed AI much closer to biology, particularly neuroscience, than it had been for many years. At the same time, as the limitations of traditional AI had become more obvious, other biologically inspired areas such as neural networks, adaptive systems, artificial evolution, and artificial life had come to the fore. From the 1990s onward these various currents mingled with the "New AI" approaches to robotics, spawning new attitudes and directions. The face of AI and intelligent robotics was radically changed. One thing all these approaches had in common was the use of decentralized, distributed, parallel processing—many things going on simultaneously—as opposed to the monumental, centralized control systems of traditional AI robotics. Parallel distributed processing promised a robust and efficient alternative to the sclerotic sequential pipeline model.

As we will see in a little more detail in Chapter 4, the New AI approaches swept to dominance and still hold sway today, although some methods now make use of updated elements of traditional AI alongside the powerful, distributed, modern adaptive techniques. The next chapter will delve deeper into the workings of modern robots. It will discuss some of the main "control architectures" used, and how the robot "nervous system" interacts with the robot body, sensors, actuators, and environment. With this deeper understanding, questions of how autonomous and intelligent robots really are can be revisited in more detail.

# 4

# INSIDE THE MACHINE

*What is actually going on? How does the robot's "brain" work?*

Through the many examples encountered so far in this book, it is clear that all robots can be described in terms of sensors, actuators, overall body parts, and a control system. Control involves generation and coordination of behavior by somehow using incoming sensory information to produce appropriate actuator outputs. The exact details of what goes on between the sensors and the actuators—how the control architecture, the robot's "brain" works—can vary enormously. In Chapter 3 we saw examples ranging from the surprisingly effective behavior of Grey Walter's tortoises, with two richly interconnected electronic valves and a few simple sensors, to the grindingly slow, brittle outcomes of the gargantuan, reasoning-based control systems of traditional AI robotics. In the former, the sensor signals flow continuously through the electronic "nervous system," directly triggering motor outputs that in turn change sensor inputs as the robots move through the environment. In the latter, sensor inputs are processed and probed to build and modify a central world model, made of logic, which is used by a series of reasoning modules to try to figure out what to do next. No module can start its work until the preceding one in the pipeline has finished. If a plan can be made, its steps are then finally translated into actuator commands, which are

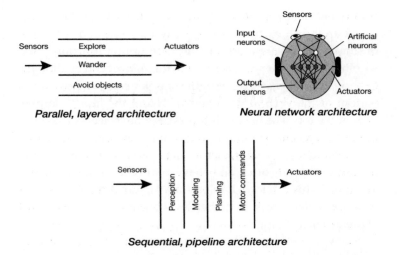

**Figure 4.1** Some standard control architecture types.

turned into motor outputs. This and some of the other main types of control architectures are illustrated in Figure 4.1.

Walter's tortoises used an early example of an artificial neural network (ANN) employed in a distributed parallel architecture. An ANN is a network of simple processing units (artificial neurons) that is very loosely analogous to a biological neural network. Later ANN-based architectures were rather different to Walter's, not least because the networks were implemented in software on computer chips. They too involved sensors feeding into designated input units and actuator signals being taken from motor output units. In between the input and output units, an inscrutable tangle of interconnected artificial neurons processed signals flowing around the network, often in rather enigmatic ways (see examples in Figures 4.1 and 4.2).

The behavior-based robotics control architectures of the 1990s and 2000s that we met in Chapter 3 were also distributed and parallel, favoring a layered design (Figure 4.1 top left), in direct contrast to the strict sequential pipeline of most traditional AI approaches (Figure 4.1 bottom). In the behavior-based approach each layer is usually made from a collection of simple processing units, or simple computer functions, that together generate a particular behavior. A layer can be thought of as a *level of competence*, with the simpler competences at the bottom of the vertical decomposition and the more complex ones at the top. The behavior layers each have access to sensor inputs and produce motor outputs—they can act as standalone control systems. They interact in such a way that, although the default position is often all behaviors running independently in parallel, this can switch to one layer overriding all others in certain circumstances, or higher-level behaviors taking control of lower-level competences.[1]

Today, as we'll see, although most practical architectures are distributed and parallel, they are often hybrid, using elements of all these approaches.

### How many ways are there to control a robot?

Countless. As many as the fevered imaginations of engineers and scientist can dream up. However, at any one time there will be just a few favored methods and techniques that most people use. These will change over time as new discoveries occur and certain methods prove themselves to be better in important practical cases. Most scientists and engineers follow the herd, but there are always a few mavericks and rebels who are crucial to pushing things forward.

Roboticists have a variety of motivations which will influence the kinds of methods they use. Some are heavily motivated by a particular theory and philosophy—they will want to demonstrate its worth and probe its limitations. Others, particularly those working in commercial environments, are much more motivated by getting stuff to work well in real applications. The latter case is generally much less purist—whatever does the job now is fine.

The overall approach taken will also be task dependent. The detailed requirements for control of a visually guided autonomous mobile robot are generally rather different from those of a bench-bound non-autonomous robot arm used to stack parts on a conveyor belt, as are those of an eight-legged walking robot moving materials around the complex terrain of a forest compared to a small flying drone intended to provide rapid active surveillance in a giant warehouse.

### How do driverless cars work?

A good way to start exploring the inner workings of modern mobile robots in more depth is to tackle one of the most important questions of the day: How do driverless cars work? Autonomous vehicles are a very good exemplar of current mobile robotics as they employ many of today's cutting-edge techniques, and indeed have been a major driver in their development. They may also have a significant impact on many lives in the near- to mid-term. So let's look at them in a bit more detail.

Driverless cars are mobile autonomous robots that use a variety of sensors to pull in information about the surroundings, and a series of actuators to control steering, acceleration, and braking. They have one main task: get from A to B without causing any danger or damage, while staying strictly within the rules of the road. In closed, quiet neighborhoods this is a relatively constrained problem. But, in general, in potentially busy, volatile traffic conditions where pedestrians and cyclists might be thrown into the mix, this is a non-trivial problem. To understand how things work today it is useful to first go back in time.

### When did work on driverless cars start?

Attempts to develop driverless vehicles go right back to the 1920s, only a few years after the first production automobiles started to appear. At that time, radio-controlled cars, which could automatically follow a leading vehicle transmitting instructions, were investigated. From the 1930s onward a great deal of effort went into systems involving cars that followed guiding circuits embedded in the road, with automatic control of traffic at junctions. While some of these approaches looked promising, it became increasingly clear that they would be difficult to implement in practice. Unless everyone was forced to use automatically guided vehicles it was hard to see how the system would work smoothly with a mix of guided and normal vehicles. And then there was the question of the cost of embedding the necessary systems in the roads. But perhaps most importantly, it was unclear how safe these vehicles would be. They were road followers, their sensors geared toward that task. What would happen if one of them broke down, or someone stumbled into their path, or something completely unexpected happened? From the 1980s onward attention switched to equipping standard vehicles with sensors and "intelligent" control systems that would allow them to operate safely in normal traffic.

A very significant step in this quest occurred in 1989 when Dean Pomerleau and a team at Carnegie Mellon University (CMU) demonstrated how an artificial neural network could be used to successfully steer a test vehicle around the CMU campus roads.[2] Pomerleau's system, ALVINN (autonomous land vehicle in a neural network), controlled the steering of CMU's NAVLAB, a converted Chevrolet van used for autonomous vehicle research.[3] NAVLAB had two main sensors, both mounted on the roof above the cab: a color video camera and a scanning laser rangefinder. It also had racks of computers and other equipment in the back which were connected to the sensors and actuators of the van (which was of course a large mobile robot). ALVINN used two "retinas" as sensory input. The first was derived from a 30×32 image continuously fed from the camera (very low resolution by today's standards) and the second was an 8×32 "image" from the laser rangefinder (which scanned over a volume 80 degrees wide by 30 degrees high). The values in this second retina were proportional to the distance to surrounding surfaces at the corresponding points in the scanned image. These surfaces could include the road, the verges, trees, or obstacles. The retinal values were fed directly to 1,216 input units (one for each retinal element) of an ANN (see Figure 4.2 for an illustration of the kind of ANN used). The output of the network determined the steering angle the wheels needed to be set at to accurately follow the road. This output was used by a servomechanism to automatically steer the vehicle. A human was in charge of other controls (accelerator, brake, etc.), but ALVINN was able to steer autonomously.

### How do the artificial neural networks used in cars and robots work?

The units (artificial neurons) in an ANN are joined together by connections through which "signals" flow (Figure 4.2). An ANN will have some dedicated input and output neurons which connect it to the task it is to perform (for a robot these will usually be directly connected to sensor inputs and motor outputs, as illustrated in Figures 4.1 and 4.2). The artificial neurons typically perform some simple mathematical transformation on the sum of all the signals from their incoming connections. This is used to produce an output value (which then becomes an input to any neurons connected downstream).[4] For instance, the neurons in the middle (hidden) layer of the three-layer network shown in Figure 4.2 each take inputs from all the neurons in the first (input) layer. They then pass their output simultaneously to each of the neurons in the third (output) layer.

Although ANNs can be built out of hardware, as Grey Walter did all those years ago, nowadays in most cases they are implemented in software; that is, they are simulated on a computer. The connections between units usually have "weights" (or strengths): these are simply numbers in some

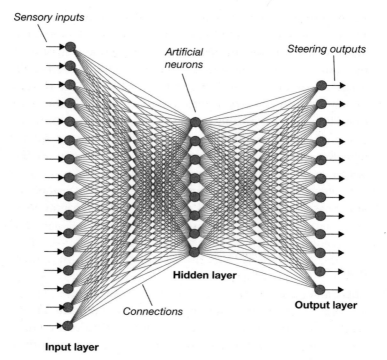

*Sensory inputs*

*Artificial neurons*

*Steering outputs*

**Hidden layer**

*Connections*

**Output layer**

**Input layer**

**Figure 4.2** An ANN of the kind used by an autonomous land vehicle in a neural network (ALVINN). Filled circles are artificial neurons which are joined together by connections, represented by lines. This is a three-layer feed-forward network. The input layer feeds into the hidden layer which feed into the output layer. Signals flow from sensory inputs through to motor outputs. Each of the neurons in the input layer connects to all the neurons in the hidden layer, and each neuron in the hidden layer connects to all the neurons in the output layer.

pre-defined range (for instance –2 to 2). The output from the neuron the connection originates *from* is simply multiplied by the connection's weight to produce the input at the neuron it is connected *to;* in other words, the signal flowing through the connection is multiplied by the weight. This means the signal on a connection with a higher weight will have more "influence" at the neuron into which it flows. In most cases an ANN "learns" by adjusting the values of these weights (increasing some and decreasing others). The input values feed through the mesh of weighted connections, combining in potentially complex ways, producing the output signals at the other end. Hence if the weights are changed, the network will process the values feeding through it differently and produce changed outputs. The idea, then, is that the weights are altered, according to the rules of a learning algorithm, until any set of inputs for the task at hand produces the desired outputs after they have flowed through the network. Learning is a matter of searching for the best set of connection weights in an efficient way.

ANNs started to become particularly popular in robotics and AI at about the time of Pomerleau's pioneering work at CMU because of a number of key properties that had potential to overcome the weaknesses of traditional AI methods. Apart from their ability to learn and adapt, and their use of parallel distributed processing, they could generalize to unseen data and deal with incomplete and/or noisy sensor signals.

There are generally three types of learning that can be applied to ANNs. *Supervised learning* involves the ANN being given many training examples where the desired output is known. After each example the weights (and sometimes other properties) of the network are adjusted according to how close the actual network output is to the correct output. As learning proceeds, the weights shift and jiggle around until they crystallize into a stable set of values. Successful learning produces a set of weights such that inputs to the network produce the desired output (in the case of ALVINN, the steering angle as in Figure 4.2).

As the name suggests, *unsupervised learning* does not make use of comparisons with the correct answers; it is not given explicit guidance. Again, the network is given many training examples, but learning now involves the ANN self-organizing around hidden patterns it detects in the data. These patterns often involve complex interactions between information from different inputs (sensors in the case of robots) which are difficult/impossible for humans to spot. This information can then be very helpful in classifying inputs into different groups. These might refer to different types of objects picked out by a robot's visual system, or the classification of different types of environmental conditions which can then be fed into a robot vehicle's control system. In unsupervised learning the ANN weights are often adjusted using a remarkably simple but effective learning rule first proposed by Donald Hebb back in 1949.[5] Using the Hebb rule, a connection between two artificial neurons that are simultaneously strongly active is strengthened; otherwise the connection weight is reduced.

*Reinforcement learning* involves using feedback from the environment to refine a behavior. It makes use of reward signals that indicated the appropriateness of a system's action in relation to its current state. This can be particularly suited to some problems in robotics. Robot actions in the environment can generate a reward related to its overall behavior (e.g., how far it moved without crashing, how many times it moved in the correct direction according to an external visual cue, and so on). The idea is to maximize reward. The reward signals help to guide a trial-and-error learning process toward good behaviors—plenty of reward means the robot is doing the right thing; little reward means it is not and needs to try something else. Popular methods at the moment include deep reinforcement learning where a deep (many-layered) neural network processes sensor input and produced motor outputs. As it proceeds, the network's weights and other parameters are adjusted by algorithms that take into account the amount of reward received.[6] Gradually the system settles down into a state where it is doing the right thing more or less all the time.

There are numerous variations on the exact details, but these kinds of processes are at the heart of nearly all modern machine learning, which now underpins much of robotics and AI. Many robot control systems use combinations of all three types of learning, distributed across various sub-systems.

## What was the structure of ALVINN's neural network?

ALVINN's ANN was a three-layer feed-forward network. Although somewhat larger, the basic structure was the same as the network shown in Figure 4.2. An input layer fed into a middle layer which fed into an output layer. The sensor values fed directly into 1,216 input units. Each of the input units was connected to every one of 29 units in the middle (so-called hidden) layer, making a total of $1,217 \times 29 = 35,293$ connections (and weights) between the first two layers. Each of the middle layer units was connected to each of the 46 units in the output layer, making another $29 \times 46 = 1,334$ connections. Forty-five of the output units represented steering angles (the unit with the highest output determined the angle to be used by NAVLAB).[7]

ALVINN's ANN was trained with recorded and simulated input data using a supervised learning method.[8] The trained network was able to successfully steer the vehicle along a 400-m course through a wooded area of the CMU campus under sunny conditions. The whole setup pushed the limits of available technology at the time (which was much slower than current computers and sensors), and so the vehicle was only able to travel at 0.5 m/s (a little over 1 mile/hour), but that was state-of-the-art in 1989.

From these beginnings—a van crawling at snail's pace along a short stretch of road on a quiet campus—the technology developed (aided by rapid increases in computer power) such that within a few years ANNs could help control vehicles traveling at normal traffic speeds on open roads. Larger networks with more layers and inputs (so-called deep learning neural networks), using extended versions of the learning algorithms employed by ALVINN, are now routinely at work on many autonomous vehicles, particularly for visual processing. They are used for recognizing pedestrians or cyclists, or interpreting road signs, as well as for keeping in lane and road following. Another useful property of neural networks, as exemplified by ALVINN, is the seamless way data from different kinds of sensors can be integrated. The camera and laser rangefinder inputs are fed into the network in exactly the same way. They mingle and combine in potentially complex ways as they feed through the network. This merging of information from different sensor types is known as *sensor fusion* and can be very useful in coping with sensor noise and uncertainty—two sources of information is usually better than one.

### How does an autonomous vehicle "know" where it is?

ALVINN set the ball rolling on one aspect of driverless vehicles—steering—but it did not produce a fully autonomous system. There were still many other parts to tackle, not least being able to plan and follow a route in order to get from A to B. This usually requires a solution to the mapping problem: building some sort of representation of the robot's environment that can be used for accurate navigation. A related problem is that of localization—the ability of a robot to determine where it is, relative to a map, from its sensor readings (commercially available GPS SATNAV is not accurate enough to do this safely; it can have errors of up to several meters, which could be disastrous for a self-driving vehicle, but it can be a useful extra input).

Since the early 1990s, following pioneering research at Oxford University, most work in this area has concentrated on the simultaneous localization and mapping (SLAM) problem.[9] This requires a mobile robot, when placed at an unknown spot in an unknown environment, to incrementally construct a consistent map of the environment, at the same time as determining its location on the map. A great deal of progress has been made on this problem, and for certain types of environments very good solutions have been found. Nearly all these solutions rely on probabilistic inference. Indeed, real progress on the problem was made only once it was cast in probabilistic terms—that is, in terms of probabilities (or likelihoods) that certain things are true (the current location is $X$, the current sensor readings imply that the location is not Y, and so on). Once these values have been estimated with sufficient accuracy, the location can be assumed to be the place with highest probability. Today's solutions use probabilistic models of the robot and its environment, employing probabilistic inference in building maps from the robot's sensor readings. More and more accurate estimates of the pertinent probabilities are built up as the robot gathers evidence about the world as it moves through it. The success of the probabilistic approach stems from the fact that the mapping problem is inherently uncertain and robot sensors are noisy, as is robot movement. Sensor readings drift and fluctuate due to inherent noise in the electronics or changes in environmental conditions. Motor actions are often not quite perfect due to inherent noise and inaccuracy in the physical systems making up the actuators, or through tricky environmental conditions (e.g., a robot foot or tire might not grip quite as well on soft ground as on firm surfaces). The probabilistic approaches embrace these characteristics of the problem rather than ignoring them or trying to hide them.

Typically a map describes the locations of and relationships between key features picked out by the robot's perceptual system, along with various aspects of the geometrical layout of the environment. The process of refining estimates of probabilities through the gathering of evidence, that is, the values of the probabilities associated with the robot position and the map, is made possible by what is known as Bayesian inference. It is named after Thomas Bayes, an eighteenth-century English mathematician and Presbyterian minister who first developed some of the underlying mathematics (known as Bayes' theorem). Well-defined, efficient mathematical procedures have been developed to make the necessary updates and refinements, ensuring that the estimates of probabilities become better and better as the robot gathers more information about the world.[10] After a while the estimates become extremely good, allowing locations to be determined very accurately. Bayesian approaches to the robotics SLAM problem emerged at about the same time as behavior-based and biologically inspired approaches. Although its methodology is mainly complementary to that of those areas, it certainly has overlapping concerns. Just as biologically inspired robotics can be traced back to the work of Grey Walter, many of the methods at the core of probabilistic robotics also have their roots in cybernetics.[11]

One of the first really notable successes using these Bayesian methods was achieved by a team from Stanford University led by Sebastian Thrun, who in 2005 won the DARPA Grand Challenge for an autonomous vehicle to navigate across an un-rehearsed 142-mile course in the Mojave Desert.[12] A few years later Thrun was brought in by Google to start a research program on driverless cars, which eventually led to the emergence of Waymo, a subsidiary company which is currently one of the global leaders in self-driving vehicle technology.

### What kind of sensors do current autonomous vehicles use?

Current leading-edge autonomous vehicles use a variety of sensors (see Figure 4.3). The main varieties are: vision

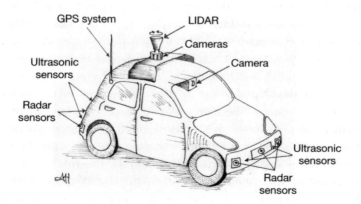

**Figure 4.3** The most common sensors used by driverless cars.

(cameras), light detection and ranging (LIDAR), radar, ultrasonics, and GPS information. Quite often, all of these are used together.

LIDAR[13] sensing involves a rapidly rotating array of laser rangefinders mounted on the roof, giving a real-time, 360-degree panoramic image of the surroundings up to a range of about 200 m. It provides very accurate information on distances to surrounding surfaces, which can be used to create a kind of 3-D model of the environment. LIDAR has been shown to be effective in building up maps of urban areas using the probabilistic SLAM approaches outlined earlier. To date, Google/Waymo have focused on an approach to autonomous vehicles that involves building very good maps of certain areas with LIDAR. Vehicles always have access to these maps such that they only need to concentrate on the localization part of the SLAM problem, again using LIDAR data. This means they can usually navigate well on networks of pre-mapped roads, which could be effective for urban driverless taxi services and the like. These maps are consulted by the control system to check on location, make decisions about when to change speed, decide if it's safe to make a turn, and so on.

Forward and rear bumper-mounted radars (which bounce radio waves off objects in their path) are able to detect the approximate speed and distance of the vehicles ahead and behind, and can locate other objects in the vehicle's path. Ultrasonic sensors (as used in automatic parking systems) detect very close objects to help avoid collisions or bumping into the curb.

Vision can be used for a variety of things including road following and identifying road signs, pedestrians, other vehicles, and hazards. Most visual systems are powered by (deep learning) ANNs which have been trained to make appropriate classifications (e.g., pedestrian ahead) or output control signals (e.g., steer slightly to the right to stay in the center of the lane). Typically, several ANNs doing different jobs will be taking inputs from the visual system.

The various sensors all have their strengths and weaknesses but work well in combination. LIDAR is fairly high resolution and gives extremely accurate information on distances to surrounding surfaces and objects (depth information), gives 360-degree coverage, and works in the dark, but it is very expensive—an automotive grade LIDAR system costs about $80,000 (in 2019). However, the price of a LIDAR unit may plummet over the next few years as companies work on developing much cheaper systems.[14] In contrast, vision sensors (cameras) are cheap and provide very high resolution—but sometimes noisy—information. Roads are designed for humans, and our primary sense for driving is vision. While ANN-powered vision modules are no match for our visual system, they are good at recognizing key features, interpreting road signs, and, as we've seen, can successfully look after road following. However, they do not work in the dark, it is difficult to retrieve depth information from them, and their performance degrades badly in poor weather conditions (fog, snow and ice, heavy rain). Radar has none of these problems and is cheap, but it is low resolution; it is mainly used for fail-safe obstacle detection and for measuring the speed of surrounding traffic.

Typical autonomous vehicle control architectures are distributed and parallel, with multiple sensor systems feeding into various AI and control modules. The AI systems often include SLAM navigation, visual recognition, road following, obstacle avoidance, and overall planning and coordination elements. These modules are spread over multiple on-board computers and special-purpose chips which must communicate and coordinate in order to coherently control acceleration, braking, and steering in order to safely achieve the current goal, which might be to stay in lane, or change lane, or turn left at the junction when the traffic lights go green, or slow down to avoid the pedestrian crossing the road, and so on.

## What are the main approaches to achieving the goal of widespread fully autonomous vehicles?

At the moment there are basically two main approaches to autonomous vehicles. The first, as exemplified by Google/Waymo and Aurora (who bought Uber's autonomous vehicle division in December 2020), is based around fully autonomous vehicles operating in slow-speed, urban areas that have been mapped in great detail and present relatively constrained, benign driving conditions. Such an approach currently relies heavily on LIDAR-based localization and navigation. As progress is made on solving all the problems associated with perceiving and correctly interpreting unexpected events in the environment (which might be rare but could lead to a fatal accident if not handled properly), the longer-term plan is to move out into less-constrained scenarios. The direction of travel is toward more and more sophisticated autonomy, with the ultimate aim of being able to handle any driving conditions, not just the simplest. This is proving to be a challenging goal.

The second approach, as exemplified by Tesla and many of the mainstream automotive manufacturers, is to move from semi-autonomous vehicles toward fully autonomous vehicles. Semi-autonomous vehicles make use of increasingly sophisticated autopilot, cruise-control-type systems in certain driving conditions. But the human driver is nearly always in the loop. Currently, the semi-autonomous systems mainly use vision, in conjunction with radar, ultrasonics, and GPS SATNAV data. They work particularly well in another kind of relatively constrained, but very different, scenario: highway/motorway driving. The human driver can take over at any point and can switch between manual driving and autopilot mode. These kinds of systems enable the vehicle to automatically steer, accelerate, and brake within its lane, and, in some more advanced systems, even change lane and overtake other cars. Descendants of ALVINN, they make heavy use of neural net-based AI and are thus able to learn and improve over time as more and more training data become available as it is collected by the vehicle and others like it. Currently they require active driver supervision at all times, except in certain restricted conditions. In the longer term the plan is to move toward full autonomy. A big difference between truly autonomous and semi-autonomous vehicles is that in the latter the driver is ultimately responsible for safety; in the former AI has our lives in its hands.

Another area where there is a great deal of research, particularly in relation to semi-autonomous cars, is in using AI for monitoring the state of the driver. Cameras and other sensors are being used in conjunction with machine-learning systems (usually ANNs) to spot if the driver seems not to be paying proper attention to the road, or is agitated, or sleepy, or distracted.[15] The car might then be primed to switch to a more autonomous mode, or to change the internal environment (temperature, lighting, sound) to wake up or calm the driver. Clearly this could have a potentially major positive impact on safety. Car monitoring systems might also learn a particular driver's "style" so that switching between auto and manual modes is as smooth as possible: the car could mimic the driver's style as it comes out of auto mode, thus making the transition seamless and easier for the driver to take over. The car might also learn to adjust its own autonomous driving style, within safety limits, to best suit the car's passengers. Or it might detect sleepy or boisterous children in the back and adjust conditions/provide distraction in that area accordingly. Features like this could make the whole in-car experience more pleasurable and might be one of the things that accelerates acceptance of autonomous and semi-autonomous vehicles.

Other types of autonomous robots, such as the patrolling security bots we met in the opening chapters, or some robot vacuum cleaners, use very similar AI control methods to those outlined above for autonomous vehicles. Because they move slowly and in very constrained environments, their control does not need to be quite as complex; as long as they avoid obstacles and slow down as they approach people and have appropriate safety cut-offs, there is unlikely to be an issue. But autonomous vehicles are potentially lethal devices; safety is paramount.

Autonomous vehicles exist and have been successfully tested over many millions of miles of driving, albeit in restricted conditions and often with human safety drivers in place. For instance, at the end of 2018 Waymo started a very limited driverless taxi service in Phoenix, Arizona.[16] But there are still many challenges to be met before autonomous vehicles become widespread, and before they can operate over all driving conditions and on most roads. But more of that later.

### Is it a good idea to copy biology?

In many cases, yes. A good example is locomotion in walking robots (which, as we've seen, can be very useful in rough terrain). In this instance there is a lot we can learn from biology. Hundreds of millions of years of evolution have provided us with many wonderful examples of how to walk efficiently and smoothly. This includes the biomechanics—superb designs for legs and joints and how to apply forces to them—as well as the neural control architectures and how these integrate with the subtleties of the biomechanics.

This kind of robot locomotion, which involves the coordination of multiple actuators (several for each leg), is not an easy matter. Traditional methods, often based around a central control system using rules governing the relative timing of individual leg movements,[17] usually result in rather unresponsive, "clunky" motion involving limited gait patterns. Insects, on the other hand, display a wide range of gaits and robust locomotion behaviors, including quickly recovering from damaged (or even lost) limbs—all this despite rather limited neural resources. From the mid-1980s onward, insects thus became an important source of inspiration for roboticists interested in walking machines. Rodney Brooks produced a distributed dynamical network controller for Ghengis, an early example of an insect-inspired walking robot.[18] Another influential strand of work originating in the late 1980s was the neural architecture introduced by Randy Beer and colleagues to control walking in a hexapod robot.[19] This wonderful example of a distributed architecture was based on studies of the neural circuits used in cockroach walking. Generalizations and extensions of it have been much used ever since.

The key lesson from biology that influenced these designs is the following. Rather than having one centralized controller that commands the legs in a master/slave fashion, animals distribute control across the physical characteristics of the legs themselves, neural circuitry local to each leg, and networks that interconnect these local leg controllers. In a distributed controller, stable, efficient gaits arise from cooperative interactions between the many different components making up the whole system.

Of course ANNs in general, including deep learning architectures—staples of many modern intelligent robot controllers—have their origins in biological inspiration.

Another interesting methodology to have come from biology is evolutionary robotics.[20] The idea was first proposed in 1950 by Alan Turing.[21] He suggested that worthwhile intelligent machines should be adaptive and should learn and develop, but conceded that designing, building, and programming such machines by hand will be difficult. He went on to sketch an alternative way of creating machines based on an artificial analog of biological evolution. Each machine would have hereditary material encoding its structure (artificial genes), mutated copies of which would form offspring machines. A selection mechanism would be used to favor better-adapted machines—in this case, those that learned to behave most intelligently. It was more than 40 years before Turing's long-forgotten suggestions became reality.

From the outset, the vast majority of work in this area, which started in the late 1980s, has involved populations of artificial genomes (sets of genes represented as lists of characters and numbers) encoding the structure and other properties of artificial neural networks that are used to control autonomous mobile robots required to carry out a particular task or to exhibit some set of behaviors. Other properties of the robot, such as sensor layout or body morphology, may also be under genetic control. The genomes are mutated and interbred, creating new generations of robots according to a Darwinian scheme in which the fittest individuals are most likely to produce offspring. Fitness is measured in terms of how well a robot behaves according to some evaluation criteria; this is usually automatically measured, but may, in the manner of eighteenth-century pig breeders, and in keeping with Turing's original proposal, be based on the experimenters' direct judgment. This method has been used to automatically design novel, often very concise, controllers for many different, but mainly quite simple, behaviors. It can also be used to fine tune and improve existing designs, which can be very fruitful. It has close parallels to other forms of machine learning such as reinforcement learning.

Hod Lipson and Jordan Pollack, working at Brandeis University, Massachusetts, pushed the idea further with ground-breaking work on fully evolvable robot hardware, where entire autonomous "creatures" (body and neural controllers) were evolved in simulation out of basic building blocks (neurons, bars, actuators). The fittest individuals were then automatically fabricated using 3-D printing, thus achieving autonomy of design and construction.[22]

As well as walking, insects have inspired other robot behaviors. In marked contrast to most current artificial systems, insects learn to visually navigate around complex environments in remarkably few trials, and use vision to perform many rapid and intricate maneuvers. Given their relatively tiny neural resources, they must make use of exquisitely clever innate behaviors crafted by evolution, along with ultra-efficient processing methods. Studies of insect behavior and their associated neural mechanisms have started to reveal some details of the cunning strategies involved, thus uncovering a potentially rich seam of inspiration for highly efficient, yet robust, robot algorithms. Such methods could be very useful for robot navigation in certain applications (e.g., agriculture) in areas where there is limited or no GPS coverage and expensive, detailed LIDAR mapping is infeasible, or in applications where computational processing/power/weight must be minimized (e.g., in miniature flying robots or space robotics).

An example is a radically different approach to navigation based on a novel model of ant navigation focusing on scene familiarity.[23] It comes from the insight that for an ant, movement and viewing direction are inextricably linked, due to constraints on head articulation relative to body, meaning that a familiar view specifies a familiar direction of movement. Since the views experienced along a habitual route will be more familiar, route navigation can be re-cast as a search for familiar views. This search can be performed with simple scanning, a behavior that ants have been observed to perform. A (learning) neural network model of this idea proved successful at explaining biological data and has now been applied to navigation in various robots, demonstrating that visually guided routes can be learned with incredibly parsimonious mechanisms that do not specify when or what to learn, nor separate routes into sequences of waypoints.

### How reliable are current robots?

Reliability is good, but not perfect. Robots developed in research labs are often a little unreliable—adjustments and repairs are occasionally needed—but that's not the point; they are there to develop and test new theories. Commercial robots, however, need to be reliable. Exactly how reliable depends on the task they are designed to perform. The odd minor blip in a robot vacuum cleaner is unlikely to be much of an issue, but a major failure in an autonomous vehicle could lead to the loss of many lives.

There are two main aspects to robot reliability. The first is reliability of the electronic and mechanical hardware, which includes the sensors and actuators, the control system electronics and computers. High-spec quality components should be used to guard against failures; this is especially important in robots that interact with, and could potentially harm, humans. The overall integrity of the hardware design must be thoroughly tested and verified. In general, electronics have become much more reliable over recent years, but they are still prone to occasional and sudden faults. Therefore safety-critical circuitry needs to include fail-safe methods such as built-in redundancy and self-checking electronics that can initiate a graceful glide into a secure mode. Such additions greatly increase the cost and complexity of the hardware and are very difficult to make completely watertight, especially for large circuits. The automotive sector already has strict regulations and international rules about reliability and safety (but still many components fail), and these will need to be tightened as autonomous vehicles arrive. But other, as yet almost unregulated, areas of robotics probably do not have the same level of awareness of the fundamental issue of electronics reliability. Many assume it is a solved problem—it is not.[24]

The second aspect is reliability and robustness of the control algorithms. Will the robot always do the right thing? Will there be circumstances in which it does something dangerous? Again, how crucial this aspect of reliability is will depend on the task. Thorough testing and analysis of the properties of the control system are major parts of the attack on this issue. This is an area where ANNs have a potential weakness. It is often difficult to work out exactly what is going on inside a trained neural network, and therefore it is hard to examine and interpret the minutiae of how certain decisions were reached. But when an autonomous car's controller does the wrong thing, we need to know why, so it can be fixed. Hence there is currently a big drive for explainable AI systems, particularly ANN-based systems. But in truth, the explainability and exhaustive testing problems are faced to some extent by all control methods. Training and adapting, which involve exposure to vast amounts of data from a huge variety of scenarios, including the difficult, unexpected, bizarre ones (so-called edge cases), are an integral part of machine-learning approaches. If there are enough data—and in the case of autonomous vehicles, Waymo, Tesla, and the rest have collected vast oceans of the stuff—this perceived weakness could possibly become a strength. These systems will probably be the most thoroughly tested—over millions and millions of hours of training and live operation—in the history of technology. Of course, the training/test data need to be of the right quality, and the verification process (testing out in live driving scenarios) must be extensive. Testing/verification cannot have hidden biases and so must be as varied and exhaustive as possible, with numerous edge cases covered. But none of this makes the explainability problem go away; that remains a major challenge for machine learning in safety-critical applications.

Commercial pressures are forcing robots to become reliable—who is going to buy something that regularly fails or doesn't perform as claimed? But complete perfection is not possible with any technology. Electronic and mechanical failures will always occur, and one-in-a-billion freak circumstances that confuse the control system, no matter how well tested and designed, are probably inevitable.

### How intelligent are they, really?

Robots are undoubtedly getting smarter. Driving a car full of people around urban streets while obeying all traffic rules and not endangering anyone—inside or outside the vehicle—is impressive, and something that most of us would agree requires a degree of intelligence. However, even control of an autonomous car is a very specialized, constrained behavior. The cases that currently work are even more constrained (limited, very heavily mapped urban areas in benign traffic and good weather conditions). The general case—of being able to autonomously drive anywhere in any conditions, with or without a detailed map—is probably still a long way off. Being able to make safe, sensible decisions while approaching a complex crowded junction full of cars and pedestrians during a driving rain storm after the traffic lights have failed, and in the midst of a full-scale riot, requires a more general kind of intelligence that can extrapolate and improvise. How long it will be—if ever—before robots are capable of such behavior is an open question. There is no evidence that it is just around the corner, as some voices, more hopeful than informed, would claim.

Self-improving, adaptive, generalizing capabilities are an important part of the higher robot intelligence we are striving for. Relevant skills include transferring something learnt in one situation to another different situation, or being able to generate novel strategies for problem solving. AI techniques for such qualities are developing, but it is impossible to say when they will be mature. It could easily be many decades. There are still quite basic issues that are very challenging (e.g., perception in foul weather for outdoor robots; decision making in unfamiliar, unstructured situations, or learning from very few examples, for all robots).

At the moment the ways in which humans can communicate with most robots are still quite limited. But AI-based voice recognition and interpretation has come a long way in the past few years (mainly thanks to machine-learning methods), and so robot voice command and response is a potentially powerful route that is being explored with some vigor. Anyone who has experience of the various "home assistant" smart speaker devices knows that parsing and comprehension of spoken language is pretty good, if far from perfect. These devices are connected to the cloud, where huge computing resources are used to power the AI behind them (including large neural networks). A similar approach would work for robots that were permanently connected to the cloud. But in some safety-critical cases where a dropped wireless connection could not be risked, a probably more limited self-contained on-board system would have to be used, and, if lives are potentially at risk, it would have to be close to perfect. Of course there are numerous other ways humans communicate, including reading body language and facial expressions, and through gestures. Research into providing robots with similar capabilities has been ongoing for some time and is making progress.[25]

So, for the moment, robot intelligence remains very narrow and very task specific. It is limited but has got to a stage where it is genuinely useful. This chapter has mainly focused on driverless cars as an exemplar of autonomous robots; we'll explore the state-of-the-art in some other areas in later chapters, including unnerving possibilities being pursued by the military. Until then it is worth noting that many of the most impressive examples developed to date have involved huge teams of engineers and vast amounts of training data, only possible through enormous corporate financial resources—but exactly how deep those pockets are is unclear. Robot AI will continue to develop, but trying to predict how fast and how far is largely futile. Anyone who tells you otherwise should be treated with great skepticism. The methods that will enable less narrow, more flexible intelligence might be very far from what we have now, may not even have been imagined yet.

But what it might be like to interact with such robotic intelligence has been imagined—in films, literature, plays, and TV shows. Depictions of robots in popular culture can be a powerful way of illustrating and thinking about the potential opportunities and dangers of robot futures. Chapter 5 begins to explore this theme.

# 5

# ROBOT FANTASIES

## ROBOTS IN POPULAR CULTURE

*When did robot-like machines first appear in literature?*

The depiction of robots in popular culture has long shaped common perceptions of what they should look like and be capable of, but has also influenced the development of actual robots. Many of today's leading AI and robotics researchers cite reading novels by writers such as Isaac Asimov and Arthur C. Clarke in their youth as a major influence on the direction they took. For almost a century there has been a twisting symbiotic relationship between robots of the imagination and machines developed in laboratories.

We have already seen that the word robot came from literature, specifically Karel Čapek's play *R.U.R.* But perhaps the first literary appearance of robot-like machines was in H.G. Wells' celebrated novel *The War of the Worlds*, originally published in 1898 and still popular today, having been the subject of numerous film, radio, and TV adaptations.[1] The story revolves around the invasion of Earth by brutal yet technologically advanced Martians in search of resources because those of Mars are almost gone. The Martians used four types of machines to aid the invasion: the fighting machine, the handling machine, the flying machine, and the embankment machine. The first of these were giant three-legged walking machines that strode across the landscape unleashing eviscerating blasts from a heat ray. They were not autonomous, being operated by a Martian pilot, but their smooth locomotion suggests that at least their walking mechanism was automatic, if guided by the pilot. The less ferocious handling machine was a spider-like crawler used to lift and manipulate heavy objects. They too were operated by a Martian controller but, from the brief description in the book, seemed to have had some level of automation. From a robotics point of view, perhaps the most interesting was the embankment machine, a digger sent out to widen the landing sites of the Martian spacecraft. The novel's narrator explains that the machine has no pilot or any form of cockpit—it appears to be autonomous.

There were multiple levels of meaning and commentary in Wells' book: on the evolution of intelligence and emotions, on the ruthless destruction wrought by British imperialism, but also on the possible influence of, and reliance on, future machines.[2] Wells was a trained scientist, having studied biology under T.H. Huxley (known as Darwin's Bulldog for his strong advocacy of evolutionary theory), who thought deeply about technology and what it might become, for better or worse.[3]

Wells' outrageously imaginative visions of walking and crawling machines influenced later generations of science fiction writers and film makers—remember the scuttling, spider-like, miniature surveillance robots in Steven Spielberg's 2002 film *Minority Report*? These fictions in turn influenced late-twentieth-century roboticists as they experimented with designs for walking machines. Wells had rightly reasoned that walking machines could be a powerfully general and adaptable way to traverse rough and varied terrains, especially if the limbs were multi-purpose and could be used to dig and push and grasp as well as balance on. This was exactly the thought behind current developments of legged robots for applications in forestry and agriculture as well as for the military.

H.G. Wells' devices were perhaps the first fictional machines of a kind we would now recognize as robots, or at least robot-like: mechanical, inorganic structures bristling with appendages, joints, and levers. A few decades later the predominant popular vision of robots was of mechanical humans built from steel and electronic circuits. Fritz Lang's 1927 silent film *Metropolis* was very influential in embedding this image. Lang's visionary film, written by his then wife Thea von Harbou, told of inequality and class struggle in the futuristic city of Metropolis, where overlords lived in luxury, while the workers were forced to exist underground, brought up only to tend the machines that created the city's wealth. Large parts of the plot revolve around a humanoid robot built by a crazed, maverick scientist. The film is highly stylized and the robot is visually stunning, all gleaming metal and futuristic design with a distinctly female form.[4] This iconic image of the machine-human, with a sleek metal body adorned with occasional ridges and protuberances, hinting at the internal mechanics, became very influential—the design of C-3PO, from *Star Wars*, is clearly strongly based on the *Metropolis* robot.

### What about organic, bio-engineered robots?

Robots as obviously inorganic machines dominated the early sci-fi pulp fiction of the late 1920s–1950s and many films from the 1950s onward (Robbie the Robot in the popular 1956 film *Forbidden Planet* is a prime example). But there was another tradition that pre-dated *The War of the Worlds* and fed into *R.U.R.*: that of organic artificial humans, somehow created through biochemical processes in laboratories. In *R.U.R.*, Čapek relied on a mysterious chemical accidently discovered by Old Rossum, a marine biologist working on an isolated island. The properties of Rossum's chemical were almost exactly the same as those of living biological cells, but it was more robust and could be spun and kneaded and molded into biological organs and bones. Old Rossum spent years trying to perfect the creation of animals to prove that God was unnecessary. His nephew, young Rossum, saw the potential for creating artificial humans—robots—which could make him enormously rich. He perfected this process, and the robots thus constructed, on production lines in factories, were not humans, although they looked just like us, but could be regarded as living beings. A cruder precedent for this idea, which no doubt inspired Čapek, was H.G. Wells' 1896 novel *The Island of Dr Moreau*, in which a mad vivisectionist, also working on an isolated island away from prying eyes, creates humanoid creatures by stitching together parts from cut-up animals. Both of these tales of course owe a debt to the granddaddy of all modern science fiction stories: Mary Shelley's *Frankenstein*, which was first published in 1818. Dr Victor Frankenstein created his humanoid monster by carefully piecing together biological material collected from dissecting rooms and slaughter-houses. He had discovered some unspecified fundamental principle of life which allowed him to reanimate the organs of his creature, imbuing it with life through a shadowy process which is never explained (the image of high-voltage electric shocks bringing the monster to life mainly came from later film adaptations).

The most extreme version of factory- or laboratory-created humanoids comes in Aldous Huxley's classic 1932 novel *Brave New World*. Here we meet not humanoid robots but mass-produced humans *as robots*. In this dystopian story, set hundreds of years in the future, babies are created in factories from harvested human ova and spermatozoa through carefully orchestrated processes of insemination and gestation in artificial wombs. Using a combination of genetic engineering and rigid conditioning, people are created in strict strata, from the alphas at the top to the epsilons at the bottom. The alphas and betas undergo "normal" development in the incubators, while fertilized eggs of the lower strata are made to bud and divide, creating up to 96 identical clones. These serfs standardized men and women—all the easier to train and control—are used as mass labor. "Ninety-six identical twins working ninety-six identical machines"[5] are kept compliant with drugs and soft power, surrounded by advertising encouraging them to live out their lives as obedient consumers.

As well as being a highly inventive writer, Aldous Huxley had a keen interest in the biological sciences, perhaps inevitable given his lineage—T.H. Huxley was his grandfather, and the leading evolutionary biologist Julian Huxley was his brother, as was Nobel Prize winning neuroscientist Andrew Huxley. When Grey Walter (creator of the autonomous tortoise robots) was a research student in neurophysiology at Cambridge, at about the time *Brave New World* appeared, Aldous came to see his experimental setup, taking great interest in the way Grey used drugs to alter the properties of the (frog) nerve fibers he was studying.[6] So, in addition to the social and political commentary, satire, and prophecy, Huxley's work was informed by, and speculated on future uses and abuses of, science. His description of the Central London Hatchery and Conditioning Centre, which takes up the first chapter of *Brave New World*, is among the most detailed and plausible visions of future technology in science fiction up to that point.

Some of the most interesting examples of organic robots can be found in the paranoid, seething novels of Philip K. Dick, a writer of unruly genius who obsessively explored identity and the nature of reality. In his 1968 novel *Do Androids Dream of Electric Sheep?*, bio-engineered humanoid robots, referred to as androids in the book, are used for labor in the colonization of other planets as humans gradually leave a devastated Earth. Factory produced, the robots have artificially implanted memories of childhoods and pasts that never happened. Almost indistinguishable from humans, some androids go rogue and escape to Earth where they are chased by bounty hunters. Ridley Scott's dazzling 1982 film adaptation *Blade Runner*, in which the androids are evocatively renamed replicants, leaves out many of the most philosophically interesting and strange elements of the book, but nevertheless helped to reinforce the organic robot as a staple of mainstream sci-fi films and TV.

Numerous fictional machines have been spawned at various points along the inorganic–organic robot axis, although often they are clustered at either end: purely inorganic or wholly organic. Hybrid robots—part inorganic (perhaps the skeletal structure and some sensors and actuators) and part integrated bio-engineered organic material (perhaps most of the nervous system and some of the musculature)—do not feature as often. The TV series *Westworld* (2016–) explores this issue a little. The "hosts," humanoid robots for the entertainment and indulgence of super-rich visitors to an adventure theme park, seem to be mainly organic, 3-D-printed from some advanced bio-material, although parts of their brain (and maybe other organs) are programmable inorganic hardware. There is mention of earlier models that were mostly inorganic: metallic mechanical components covered in synthetic skin; there then seems to have been a progression toward the organic end of the scale, apparently because such robots turned out to be cheaper and easier to manufacture in this future world.

It is possible that hybrid inorganic–organic robots might be the most plausible far-far-future root to humanoid (or other) robots with notable general intelligence. I speculate that the brain of such devices would be more organic than inorganic. The only example we have of general intelligence is biological: ourselves and other higher animals. Biological nervous systems are astonishingly complex electro-chemical devices, as yet little understood. Humans may never gain a full comprehension, may never be able to abstract sufficiently powerful principles of brain operation to be able to program artificial versions, but in some distant, possibly sinister, future we might be able to bio-engineer and grow organic neural tissue to be integrated with programmable hardware in the creation of robot nervous systems. The first steps toward hybrid inorganic–organic systems are already being taken. For instance, scientists at Carnegie Mellon University have recently shown how it is possible to interface synthetic biological cells simultaneously to both the environment and the internal electronics of a robot, allowing signals to pass between the environment, the cells, and the robot mechanisms.[7] They synthesized a soft robotics gripper that used bio-engineered bacteria for detecting chemicals in the environment; biological signals from the bacteria were converted into electronic signals which were used to control the movements of the gripper. The robot thus constructed was able to detect objects in its environment, via the bacterial sensors, and respond appropriately via its actuators. It is a long way from there to bio-engineered neural control systems, but the path has been opened.

Another possibility is entirely organic, but artificial, human-created creatures, probably of a much simpler kind than intelligent humanoids. A very recent, exciting development, which marks a first push in this direction, was the creation in 2019 of simple organic robots using a process of artificial evolution.[8] Dubbed xenobots by their creators at the University of Vermont and Tufts University, these tiny biological machines were first designed in computer simulations using the techniques of evolutionary robotics we met in Chapter 4. The xenobots were assemblies of passive and contractile biological cells (the latter can spontaneously contract and relax, that is pulse) which were evolved to perform some simple behavior (e.g., move forward: the pulsing contractile cells could be exploited to power locomotion). The best designs evolved in simulation were then created in a biology lab from real cells—passive skin cells and contractile heart cells—developed from frog stem cells. The newly assembled organic robots were able to perform the desired behaviors when placed in a petri dish. In the words of the scientists who developed them, the creations are novel living machines, programmable organisms, and entirely new lifeforms. It is quite plausible that this kind of technique could be used to develop useful nano robots in the future, maybe to deliver drugs after being injected into the bloodstream. Scaling to larger, more complex creatures will be very challenging, but might be possible in the long term.

### How plausible are mainstream fictional portrayals of robots?

Robot portrayals in books and films are mainly fantastical in relation to where real robotic technology is today. But that is to be expected—these are works of entertainment after all, often set in some unspecified point in the future. Fictional robots are usually much more intelligent and capable than anything we could build with current technology. Indeed, they are usually far over the horizons of present research.

Where films or books are set at some specified time in the near future, the predicted state of robotics is nearly always highly implausible. *Do Androids Dream of Electric Sheep?* was set in 1992 when first published, then 2021 in later editions; *Blade Runner* was set in 2019. Over-optimistic predictions about technological progress were commonplace in the 1950s and 1960s, but even so, the development of intelligent bio-engineered humanoid robots within 30 years of the book being written was a bit of a stretch. Aldous Huxley was wise to set *Brave New World* 600 years in the future, making the scientific progress needed for his vision more plausible. Likewise, Arthur C. Clarke, one of the most influential of all science-fiction writers, and a trained scientist who cared about the plausibility of his fictional technology, played it safe at the start of his career. In the 1940s he set his stories, which often featured intelligent robots, billions of years in the future. On that timescale any speculation might be reasonable.[9] By the 1960s he was merely projecting forward to the twenty-first century (most famously in the screenplay for Stanley Kubrick's masterpiece, *2001: A Space Odyssey*). In the latter case he vastly overestimated the advances that would be made in AI over the 40 or 50 years from when he was writing. This was perhaps understandable, given how young the field was at that time, and how enthusiastically confident were many of its leading lights.

However, one near-future/alternative-present science-fiction TV show, Charlie Brooker's brilliant *Black Mirror*, has a very good example of disturbingly plausible future robot technology. The taut, minimalist plot of one episode, *Metalhead*, involves the protagonist (Bella) being relentlessly chased by a killer robot dog. There is no back-story, but we appear to be in a post-apocalyptic world where robot guard dogs—perhaps military in origin, perhaps from some powerful corporation that once ruled—are now operating outside human control, apparently just inexorably following their programming to hunt down and destroy intruders. Obviously, similar plot elements have been used before in different contexts (the *Terminator* film franchise comes to mind), but what makes *Metalhead* almost unique is the way current-day robotics is clearly directly used as the inspiration for the fictional machines, which were in fact created using computer-generated imagery (CGI) in post-production. The look and feel of the robots and the way they move are heavily based on that of advanced dog-like robots built by Boston Dynamics.[10] When we see the world through the robot's eyes it is via (real) LIDAR images. The main behavior of the robots, to follow shrapnel-embedded trackers they fire into their prey, and the way they move over rough terrain, stopping to recharge when their power runs out, are all fairly plausibly near-future scenarios, relative to current technology (the scenes where the robot dog drives a van, trying to force Bella off the road, not so much). This plausibility was one of the things that, at least to me, made the episode particularly disturbing.

The imagination can run free in written fictional accounts of robots—and many weird and wonderful machines have graced the pages of novels since the nineteenth century—but in films and TV, budgets and available special effects have always been limitations. In the early days of sci-fi films most robots were played by a man in a costume, often as crude as a deep-sea diving suit sprayed silver, or were cheap-looking static props with a few flashing lights. Organic humanoid robots became popular because there wasn't even a need for a special costume; often some tinted contact lenses or a bit of makeup and a funny haircut did the trick. The original 1960s series of *Star*

*Trek* worked wonders with limited resources, exploring many philosophical issues relating to AI and robotics in an imaginative way. Occasionally real robots were used: Elektro the Moto-Man, who we met in Chapter 3, featured as Sam Thinko in an adult cut of the low-budget 1960 film *Sex Kittens Go to College*, where he was surrounded by striptease artists. It was probably not his finest hour. In more recent decades, as budgets increased and special effects became more and more sophisticated, so robots in films have become steadily more elaborate. With today's CGI techniques, where almost anything is possible, the latest generation of fictional machines are often beautifully and creatively designed (think of *I, Robot* and *Ex Machina*).

We shouldn't forget video and computer games. They too use increasingly sophisticated CGI animation, and sometimes even virtual robotic AI control systems to power some of the characters. For instance, the immersive open-world game *Cyberpunk 2077*[11] features an array of fairly plausible specialist robots that seem to be partly influenced by the designs and capabilities of current technology extrapolated a little into the future.

The interplay between real and fictional robots is probably stronger than ever. As far as physical design is concerned, cycles of influence flow from robotics research to film and game designers to product designers in robot companies and back again. The kinds of robot capabilities and uses dreamt up by novelists and screenwriters continue to influence researchers trying to push the boundaries of what is currently possible. As real robots become more sophisticated, they spark new ideas in the imaginations of writers and directors, and so the cycle continues.

In most cases, the question of robotic plausibility within a specific book or film or other work of entertainment is neither here nor there; we are perfectly happy to suspend our disbelief as long as the plot is exciting, or intriguing, or provocative enough. Often we appreciate that there are multiple levels of allegory and commentary—the robots are often story-telling devices to illuminate our attitudes toward each other. But there is one case where we should care. Over the past 20 years or so there has been increasing slippage between science fiction and science fact as a certain kind of, often self-appointed or media-appointed, "expert" takes chunks of science-fiction plots and pedals them as reality. Favourite myths spouted by such people include: within a couple of decades robots will be as intelligent as humans; robots are about to surpass human intelligence, take over the world, and enslave us; within a few years we will be able to upload our consciousness into computers/robots (and so on)—all utter nonsense.

Some of the less alarming of these kinds of pronouncements are sometimes born out of naïve over-enthusiasm, which is forgivable—naïve optimism is an important engine of progress. Sometimes they grow from genuine concerns that are not fully informed by knowledge of AI and robotics. But sometimes such statements are much more cynical: publicity craving hucksters and charlatans manipulating gullible or sensation-hungry media outlets in order to raise their own profile, expose their "brand," or bolster their bottom line. Some of these people were preaching the same thing in the 1980s and 1990s, claiming incredible and frightening advances by 2020—none of which has come to pass. They now move the timeline along a bit but still offer no credible evidence of the kind of progress needed to make their claims even remotely plausible. I have heard plenty of people make such assertions over the years; the vast majority of them had no, or only very tenuous, out-of-date, links with hands-on work in AI or robotics. The truth is, go to any of the leading AI, machine learning, or robotics conferences, and you'll hear keynote talks from highly respected experts telling of exciting advances, but stressing how very far from general human-level performance we are.[12]

### Can imaginary robots help us to think about the potential implications of technology?

While governments and academic think-tanks have recently started to worry about ethical and social implications of the widespread use of robots, for many decades popular culture has provided a laboratory of the mind to explore these issues. Numerous literary works, films, and TV series have illustrated, animated, and expanded on what-if thought experiments in creative and provocative ways.

Isaac Asimov was one of the first writers, in the 1940s, to produce stories that seriously examined social and ethical issues that might arise if intelligent, mechanical robots were integrated into human society.[13] His famous Three Laws of Robotics, describing ideal robot conduct (of which more in Chapter 8), were a result of this exploration. Even Grey Walter, developer of the first autonomous robots, wrote a 1956 science-fiction novel, *Further Outlook*, which imagined the implications of future technology.[14]

Alex Garland's haunting 2014 film *Ex Machina* is a good example of more recent fictional investigations of related problems. With echoes of *Frankenstein* and *R.U.R.*, the story grapples with a number of complex interlinked issues that might arise if super-intelligent robots were ever created. In a tale of rampant ego, corrupting vanity, manipulation, insanity, and abuse, the vile genius CEO of a mega tech company invites a gullible junior programmer to his remote island hideaway to perform a kind of Turing test[15] on his latest robotic creation, Ava. After spending some time with the robot, will he judge it genuinely capable of thought and consciousness, despite knowing it is artificial? As the story unfolds it becomes clear that most of the main characters' motives are much murkier than they first appeared. One of the questions Garland ponders is whether or not human-level intelligence requires the ability to manipulate and deceive.

There are numerous vivid explorations of future AI and robotic technology in popular culture, from books and films to graphic novels, and open-ended computer games that branch into multiple possibilities, allowing players to explicitly explore the ethical implications of robotic futures. But here I will concentrate on the particularly haunting meditations on AI in one of the most influential of all science-fiction films, Stanley Kubrick's 1968 classic *2001: A Space Odyssey*, one of the inspirations for *Ex Machina*.

Major strands of *2001*'s plot, written by Kubrick and Arthur C. Clarke, pivot on robot behavior. "What?" you cry, "But there are no robots in that film." Some may find this a little controversial, but I claim that just as there are robots on our roads, built by Waymo and Tesla and the rest, there was one very big robot in *2001: A Space Odyssey*—the spacecraft, Discovery One, which was bound for Jupiter. While some of its functions could be manually controlled and/or overridden, the spacecraft was essentially autonomous. Its highly advanced AI control systems ran on a specially designed computer, the HAL 9000—known to the crew as HAL—which could interact through spoken conversation. HAL monitored the internal and external environments via sensors, including slightly creepy cameras that glowed red, and could initiate numerous actions, within and without the craft, via actuators—not least keeping Discovery One on track during its long voyage to Jupiter. HAL spoke to the crew in an inhumanly calm, somewhat detached voice—a mainly reassuring tone, almost friendly, but undercut with hints of something sinister. By my definition in Chapter 2, the spacecraft was a (people-carrying) robot controlled by very sophisticated AI.

Three members of the crew, all scientists, are in hibernation for the trip, their vital functions carefully observed and maintained by HAL. Two other crew, also scientists, Dr. Dave Bowman and Dr. Frank Poole, are awake and performing their duties, largely under the supervision of HAL. We are introduced to HAL via the clever device of a BBC news report being watched by Dave and Frank as they eat their unappetizing-looking space rations. The AI system is described as the latest generation of intelligent machinery, able to reproduce most of the activities and functions of the human brain "but with incalculably greater speed and reliability." HAL is the "brains and central nervous system"[16] of the craft, "controlling all aspects" of it. In response to a question from the BBC reporter, HAL describes himself as being "foolproof and incapable of error," and also a "conscious entity."[17]

In the interview, Dave is asked about HAL's apparent emotions. He explains that HAL is programmed that way, to make him easier to get on with, "but whether or not he has real emotions, is impossible to say." HAL has a very definite personality and soon emerges as the central character of the film. In contrast, Dave and Frank are portrayed as automaton-like, with almost blank personalities. HAL professes to enjoy working in collaboration with humans, but Dave and Frank seem little more than appendages to his will.

As the voyage proceeds HAL becomes increasingly troubled about certain unusual aspects of the mission. In particular he has to keep the true objective of the voyage secret from the crew. HAL reports a problem with an antenna control device that he predicts will fail in a few days. The two scientists cannot find anything wrong with it. Fearing that the crew are losing faith in him, HAL sabotages Frank and Dave while they are outside Discovery One. Overtaken by a desire to honor the primacy of the mission, along with a strong urge for self-preservation, HAL has some kind of breakdown and becomes a murderous psychopath.

Dave manages to get back inside the craft and climbs inside HAL's "brain." In one of the most powerful scenes in the film, which gives us the iconic image of columns of red light reflected across the astronaut's visor, Dave begins methodically disconnecting circuit after circuit. As soon as HAL realizes what is happening he apologizes for his behavior and protests that he is now feeling much better. As his brain is gradually dismantled, an apparently repentant HAL tells Dave he is afraid and can feel his mind going.

By explicitly referring to HAL's emotional states, and contrasting them with the astronauts', Kubrick and Clarke were posing a series of profound questions. Can a machine have real emotions, or just simulate them? Does human-level intelligence require emotions? Can a machine without emotion ever properly understand humans or interpret their behavior? Can a machine empathize?

The exact reason for HAL's mental breakdown is left ambiguous. In the later book version of 2001, Clarke traces it to the conflict between HAL's enforced deception of the crew in relation to the true purpose of the mission and his core directive to "accurately process information without distortion or concealment." In a 1969 interview, Kubrick puts it down to HAL not being able to deal with the realization that he was fallible;[18] he really had made a mistake in diagnosing the antenna unit. Either way, we are faced with the question: what would happen if an intelligent machine with emotions and complex mental states became psychotic? In the decades since the film was made evidence has emerged to suggest that parts of the human brain and nervous system habitually operate close to the edge of stability. Indeed, it is possible that some of the incredible power of these organs is gained through a flirtation with chaos, a kind of constant restlessness.[19] Hence, a related question is: would human-level machine intelligence, complete with consciousness and emotions, require a degree of instability and the risk of mental illness? (This question is to some extent pioneered in Asimov's *Robot* stories, where a lead human character is a robopsychologist, and more recently touched on in Cary Joji Fukunaga's 2018 Netflix series *Maniac*.)

When Dave disconnects HAL he is effectively killing a conscious entity. Kubrick expertly manipulates the scene so that we begin to empathize with the frightened and apparently repentant HAL, who now seems more human than the cold, efficient astronauts.

The film is both a hymn to technology but also a warning. Kubrick's and Clarke's prophecies about how AI would advance turned out to be very wide of the mark. Twenty years on from 2001 and we're still absolutely nowhere near anything that even vaguely resembles a conscious AI—we don't even know how to define such a thing. But the world they created is so coherent, so immaculately designed and considered, that it has a kind of self-contained realism that makes its treatment of all these issues truly thought provoking.

Simmering under all this is the dark question at the heart of much dystopian science fiction: if we ever succeed in developing machines with human-like levels of intelligence and self-awareness, built to integrate with human society, will they inevitably take on our worst characteristics—violence, vengefulness, coercion, greed, and hatred?

# 6

# INTELLIGENCE, SUPER-INTELLIGENCE, AND CYBORGS

*There has been a lot of talk about the singularity in relation to robotics, but what is it?*

The notion of the technological singularity can be traced back to John von Neumann, a highly influential twentieth-century mathematician and computing pioneer. His theoretical speculation, made informally to colleagues, probably in the early 1950s, went as follows. As progress in technology accelerates, there may come a point at some time in the future when such relentless advances force a fundamental change in the way human societies function; the old ways would no longer make sense, human life would be forever altered.[1] He referred to this moment as a kind of singularity—the mathematical term for a point at which an object is not properly defined, or ceases to behave by the normal rules.

The version of this idea that is most relevant here relates to machine intelligence, and is often simply referred to as "the singularity." It was first articulated by Jack Good in the 1960s. Jack was a leading statistician with a keen interest in machine intelligence who worked closely with Alan Turing on code cracking during the Second World War. He was a member, with Turing and Grey Walter, of the Ratio Club, which we met in Chapter 3. A philosophically inclined mathematician, Jack liked thought experiments,

speculations, and extended musings. In 1962, drawing together contributions from many leading scientists and
thinkers of the day, he published the self-explanatorily ti-
tled *The Scientist Speculates: An Anthology of Partly-Baked
Ideas.*[2] One of Jack's own articles in this wonderfully quirky
collection pondered the possibility of an intelligent machine
that could be educated and taught about its own design and
could then be asked to design "a far more economical and
larger machine."[3] At this point, or so Jack claimed, there
would be an explosion in technological and scientific development as the new machine (designed by its predecessor)
surpasses human capabilities and starts to solve all the hard
problems that had evaded human ingenuity.

Jack expanded this idea of an artificial intelligence explosion in a more detailed paper written in 1963.[4] The nub of his
argument is captured in the following quote from the paper:

> Let an ultraintelligent machine be defined as a machine
> that can far surpass all the intellectual activities of any
> man however clever. Since the design of machines is one
> of these intellectual activities, an ultraintelligent ma-
> chine could design even better machines; there would
> then unquestionably be an "intelligence explosion", and
> the intelligence of man would be left far behind. Thus
> the first ultraintelligent machine is the *last* invention that
> man need ever make, provided that the machine is docile
> enough to tell us how to keep it under control. It is cu-
> rious that this point is made so seldom outside of science
> fiction. It is sometimes worthwhile to take science fiction
> seriously.

The point at which Jack's intelligence explosion occurs is what
is usually meant now by the singularity.[5]

Perhaps inevitably, Jack served as a consultant to Stanley
Kubrick while he was developing his ideas for *2001: A Space*

*Odyssey.* Some of Jack's speculations on the intelligence explosion are embodied in the character of HAL. (Jack liked to tell the tale of how he requested a fee of precisely £2,001 for this work.[6])

### Is the singularity near?

Contrary to the claims of a few so-called futurologists, speculators, and provocateurs, who warn that it is an imminent existential threat, there is no convincing evidence at all that the singularity is near (and by near we generally mean 20–40 years). Progress in AI and robotics has been steady, with some bursts of accelerated developments, but the distance from where we are now to the capabilities needed for human-level intelligence—never mind super-human intelligence—is still enormous. There are huge gaps in our understanding of how to develop more general, adaptive machine intelligence, even at a level far below that of humans. Of course it is impossible to know what developments might occur over the coming years, let alone the coming decades, and some huge unforeseen breakthrough might be lurking around the corner. But we can look back at progress over the past 70 or 80 years, examine the current state-of-the-art, and survey present research trends. What this tells the vast majority of AI and robotics researchers and engineers is quite simple: the singularity is not near. Of course that doesn't rule it out as a theoretical possibility at some unspecified point in the future. But just because we can talk about it and vaguely define it does not mean it will ever happen.

### Should we be worried?

About the proximity of the singularity, no. However, the singularity is an interesting theoretical possibility and, however distant, it is wise to consider it. But as we'll see in Chapter 8, there are social, ethical, and safety issues we should worry about that stem from the far-less-intelligent robots that will continue to integrate with human life over the coming decades.

### Are super-intelligent robots inevitable?

In my view, no. Or at least there is no glimmer of them on any visible horizon. We might develop them in the longer term, but at the moment there is no evidence that we will be able to create the kind of general AI technology that would be needed to underpin such robots.

Singularity evangelists tend to point to two possibilities for creating super-human AI. The first relies on rapid increases in the processing power of computing technology.[7] The second possibility is the successful reverse-engineering of the human brain.[8] The exponential technological improvement argument relies on observations such as Moore's Law—which roughly states that the number of transistors in a dense integrated circuit doubles about every two years, with direct implications for a computer processor's speed and size—together with a very crude comparison between digital computers and brains. The claim goes something like this: by such and such a date, the number and speed of processing elements in a computer, along with its memory capacity and data handling bandwidth—its "processing power"—will somehow match that of the brain. The properties of the brain considered in estimating its "processing power" are usually the number of neurons and connections between them and their typical speed of operation.

In my view, this line of reasoning is almost meaningless. To equate neurons and dendrites with transistors and wires, and to compare neural processing with the operations of a digital computer, makes little sense. Neurons are far more complex than transistors, capable of much more subtle and flexible operation. But the real power of the nervous system comes from its organization: the interacting networks that can reconfigure and take on multiple roles, that can grow and change, and that can adapt and learn under the influence of many different mechanisms of plasticity (by which neurobiologists mean the ability to alter or transform). Neural systems are extremely complex, with many interacting electrochemical processes at play on a whole variety of spatial and temporal scales. To imagine that the kind of processing undertaken by the brain, an exquisite product of millions of years of evolution, should directly translate to the relatively simple, basic operations of a digital computer seems very odd and belies an overly technocentric view of the world.

Jack Good was the first to be guilty of such a simplistic comparison. In his 1962 paper he estimated that his intelligence explosion would occur sometime around 1978 (although he did heavily caveat this claim). As Steven Pinker has put it, "sheer processing power is not a pixie dust that magically solves all your problems."[9] More powerful computers will help, and indeed are helping, to advance AI and robotics, but the key breakthroughs will be in developing new principles and mechanisms that underlie the generation of intelligent behavior. That is no easy matter. A super-powerful computer is just a pile of electronic junk unless it is running algorithms. Developing new kinds of revolutionary algorithms that will create a leap forward is a much more uncertain and long-term endeavor . . . which brings us to the reverse-engineering argument.

Good's second (1963) paper on the intelligence explosion opens with a statement that in order to develop human-level AI systems we need to understand much more about the workings of biological brains. I think this is still true today. The nervous system, which includes the brain, is the only example we have of a system that can produce sophisticated intelligent behavior, including advanced cognition. It also seems to be frighteningly complex, probably the most complex system we are aware of.

Claims that the brain will be reverse-engineered in a couple of decades, such that detailed computational facsimiles will run on super-computers, seem hopelessly optimistic to me. There are a few projects trying to build large-scale brain models. While they are making some progress, they are very far from producing anything that could form the basis of an advanced AI. Those who claim this approach could trigger the singularity in the near future are vastly underestimating the difficulty of understanding the brain. My own research straddles AI, robotics, and neuroscience, and thus I have worked closely with experimental neuroscientists for decades. I have discussed the reverse-engineering claims with many brain scientists and I have not come across a single one who believes they are credible; most find them laughable. Many advances have been made in neuroscience over the past hundred years or so, but there is still no accepted overarching theory of brain function, and no real deep understanding of how the organ works. Many neuroscientists feel we are still scratching the surface. Experimental investigations of detailed mechanisms and how they operate over behavior-generating networks are incredibly difficult. It will be a long time, if ever, before our understanding is advanced enough to enable detailed reverse-engineering. Even then, it is unclear if truly brain-like operations at the degree of detail needed for human-level intelligence could be implemented in a non-biological medium (i.e., an electronic computer). I believe that trying to abstract principles and mechanisms, rather than attempting to reverse-engineer a faithful copy, would be a far more fruitful direction, although also very challenging.

In 2002, a colleague, Owen Holland, and I asked Jack Good about his early speculations on ultra-intelligent machines. He agreed that his initial estimate of when an intelligence explosion would occur was so inaccurate because, like many AI researchers at the time, he completely misjudged the difficulty of creating general AI.[10] Some people are making the same mistake today.

I leave the last word on all this to the late John Holland, a renowned neural networks and machine-learning pioneer and inventor of genetic algorithms, whose influential career stretched back to the height of cybernetics in the 1940s.

> I become very cautious when I hear people claiming they are going to use evolution and they're going to download human brains into computers within twenty years. That seems to me to be at least as far-fetched as some of the early claims in AI. There are many rungs to that ladder and each of them looks pretty shaky![11]

### What about cyborgs? Will we be able to enhance ourselves robotically?

If we define a cyborg as part human, part machine, then steps in this direction have been underway for some time. Many of the most advanced applications in this area are medical and involve electronic sensor and actuator technologies to restore lost sensory or motor functions. Robotic technologies are being integrated with human bodies to replace lost limbs and partially recover senses such as vision or hearing.

For centuries, we have augmented ourselves with technology as our bodies fail or are damaged—from walking sticks, to wooden legs, to eyeglasses, to artificial limbs, to pacemakers, to cochlear implants and artificial retinas. The current generation of robotic enhancements is part of a long line of developments. Some thinkers, notably Andy Clark, have argued that when we use technology as an external cognitive tool, not necessarily directly integrated into our bodies, we become part of a cyborg system.[12] This might be as simple as writing with pen on paper. Clark argues that we offload part of the cognitive process to the technology and in doing so become part of an interacting human–technology hybrid system. By this broadened definition, we are all already cyborgs.

Some of the most advanced artificial arms and legs available now are decidedly robotic.[13] Sensors coupled with AI are used to pick up patterns of muscle movements in the amputee's stump, which are then translated into signals that generate the appropriate movement of the robotic joints. On-board computer chips provide adaptive control that aims to learn the most natural movement/gait to suit the user. Some experimental robotic limbs are connected to the user's nervous system by attaching wires directly to nerves that were used to control the original limb, before amputation. These nerves are of course already connected to the motor control areas of the brain. The idea is that the user can learn to control the artificial limb in a very natural way (essentially by thought, by willing it to move). The communication bandwidth is lower than for a biological limb, with reduced subtlety of movement, but this is a promising direction that should continue to develop.

In parallel to the developments with robotic limbs, recently there have been exciting advances in artificial retina implants. The retina is a layer of neural tissue at the back of the eye that includes light receptors. Its job is to convert visual signals into electrical impulses that are passed down the optic nerve to the visual processing centers of the brain. Retinal damage or disease can result in blindness. To date, artificial retinas consist of electronic chips that are placed at the back of the eye. An array of tiny electrodes makes contact with any remaining healthy retinal tissue. The array receives a signal from a minia-ture wireless camera that can be mounted on a pair of glasses. Images from the camera are converted into electrical signals that are fed via the array to the retinal tissue and thus to the optic nerve. Such a setup has been able to restore partial, low-resolution vision to some patients. The latest developments in-volve ultra-thin, flexible, 2-D, graphene-based retinas,[14] which promise a more bio-compatible solution that in the longer term might be able to restore greater areas of vision. The kind of processing involved in the artificial versions is far from the so-phisticated operation of a healthy retina, but it is a very prom-ising stride in the right direction.

Brain computer interfaces (BCIs) is another area of great interest in the context of cyborgs. The field was pioneered in the 1960s by the ever-ingenious Grey Walter. Grey and his team used a computer to analyze in real time EEG brain waves recorded from electrodes placed on the scalp. They demonstrated how a subject could learn to generate different EEG patterns by thought alone, which could be used to operate an external device hooked up to the computer. Grey was able to turn a TV on and off in this way.[15] The original experiment involved a large room stuffed with electronics; today, similar results can be achieved with a single EEG sensor cap, a small interfacing unit, and a laptop. Such BCIs have thus become practical and can be used to move a pointer around a computer screen, or even control robotic prostheses or wheelchairs. Some BCIs use implanted micro-electrode arrays which can read activity in specific targeted brain areas. BCI technology is still relatively crude, but it is advancing and may yet transform the lives of some severely disabled people.

Robotic exoskeletons—powered support mechanisms worn around the body—can provide helpful boosts in mobility and strength to some individuals, and can be very useful in rehabilitation, for instance when patients are learning to walk again after a serious accident. Commercial exoskeleton suits are starting to become available, mainly targeted at increasing the efficiency and reducing the stress of certain types of manual activity (e.g., firefighters climbing long ladders while carrying heavy loads).[16]

Some of the most interesting explorations of cyborg technology have come from artists. In 2007, Stelarc, who has a long history of artistic investigations of the body and its interactions with machinery, had a third ear surgically attached to his arm. The ear was partly sculpted from his flesh, part artificial scaffold, and part grown from Stelarc's stem cells. The idea was to later insert a microphone that would broadcast over the internet. A few years earlier, Neil Harbisson effectively created an extra sense for himself—or at least extended and repurposed existing senses. He did this by having a device implanted in his skull that converted electromagnetic radiation from outside the human visual spectrum into vibrations he could feel in his head and hear as the vibrations traveled to his ears. The device also responded to electromagnetic radiation in the visible spectrum, allowing Harbisson to "hear" colors. This was particularly enriching as he had been born with total color blindness, seeing only in grayscale. After a while, he adapted to the new sense and it became a natural part of him, to the extent that his dreams were suffused with the color soundscapes.[17]

### Should we be concerned about robotic cyborg enhancements?

At the moment, not particularly. Nearly all cyborg applications are medical, of a positive nature, and properly regulated. But if the technology starts being used solely to enhance the powers of healthy humans, we need to be very careful how we control it. The military have been sponsoring research in this direction for a while, mainly into the use of robotic exoskeletons to increase the power and efficiency of soldiers on difficult terrain. To date, such work has had mixed results, but it does raise worrying visions of RoboCop-type scenarios.

Some transhumanists and singularity devotees advocate the development of sensory and cognitive brain implants that would greatly boost mental abilities, adding memory stores, direct "thought-access" to high-powered computing engines, and new kinds of sensory inputs. At the moment, this is nothing more than wild speculation and wishful thinking. The brain is an extraordinarily plastic organ, but within limits. It is not a modular computer into which new units can be slotted willy-nilly. The nervous system has evolved within a very specific context: our bodies and our senses. It is not at all clear how it would react if it was somehow connected to new, completely unnatural streams of data. The brain takes a very active role in constructing our conscious experience. When senses are corrupted or damaged or the chemical parameters of the nervous system are thrown out of kilter (e.g., through the use of drugs) the natural mode for the brain is to attempt to anticipate, "fill in," and extrapolate from within the bounds of "normal" sensory reality. This usually leads to garbled sensory experiences, such as hallucinations. So far, brain implants have involved limited, low-bandwidth inputs close to natural sensory data. The addition of high-bandwidth artificial implants and unnatural inputs into such an exquisitely honed and finely balanced system is likely to be highly problematic on many levels.

Implants that could support high-bandwidth two-way flows of information, in and out the brain, could cause all kinds of ethical nightmares, especially if they could be accessed remotely by someone other than the person whose brain they were attached to. At the moment, such devices are still firmly in the realms of science fantasy, and are not going to appear in the near future.

However, as cyborg technologies slowly advance, it is certain that regulation must increase to head off unethical abuses of developments primarily intended to restore function rather than enhance natural abilities. This is not to say that in the future the human species might not agree on ethical ways to extend itself through robotic technology.

# 7

# ROBOTS AT WORK

*What work do robots do now?*

Throughout this book there have been many mentions of the kinds of jobs robots already do, from heavy industrial automation to precisions tools to aid surgeons. This chapter will look in more detail at some of the most important and interesting examples, including newly emerging applications that are likely to have considerable impact.

Mobile autonomous, or semi-autonomous, robots are a particularly important class of devices which are destined to become much more widespread. Driverless vehicles, security robots, and autonomous agricultural robots have all been mentioned, and of course autonomous vacuum cleaners are already the most popular kind of robot on the planet.

The huge warehouses at the heart of online retail operations are gradually losing their human staff as they become increasingly populated with robots. The handy thing about a warehouse environment is that it can be designed to suit the technology. The mobile robots that move stock around warehouses for companies such as Amazon or Alibaba do not have to be super smart because the warehouse is designed around a strict grid system. Fairly small, yet powerful, disc-like robots lift stacks of goods that are moved around the grid. Human pickers take items from the stacks of baskets to make up orders that are sent for packing. Markings on the floor help to keep the robots on track, and centralized systems—a bit like air traffic control systems—can choreograph their routes. On-board sensors keep the robots located and safeguard against collisions.[1]

Ocado, a British online food retailer, takes this idea to the extreme in its latest warehouse. The colossal barn-like structure is built around a 3-D grid of cells, on top of which thousands of simple robots move back and forth on rails. In the cells are stacks of crates full of produce. A robot can grab a crate, pull it up into its hollow insides, and move it elsewhere on the grid, or to the human pickers making up the grocery orders. Or the robot can drop a crate from its insides into one of the cells. Centralized control systems optimize the flow of robots and keep the stacks in the best order to allow quick access to the most popular items.[2] The intelligence of the system mainly resides in the algorithms that are designed to dynamically maximize the flow of goods to the pickers: that is, to maximize the throughput of the whole operation. Indeed, such warehouses, with their untiring robots dancing back and forth across the grid as they relentlessly move stock, are about 70% more efficient than the human-based systems they are starting to replace. The robots are electric and periodically retire to recharging bays when their batteries run low, to be seamlessly replaced by other freshly re-energized robots, so that the dance can continue uninterrupted.

One thing robots still have some difficulty with in these warehouses is the picking: taking items—which come in all shapes and sizes and degrees of fragility—from the crates or stacks without dropping or squashing them. However, there is considerable effort going into developing robot arm systems that can do this job. With multiple cameras and on-board neural network-based AI to control coordination and spot the right item to pick, these otherwise fairly conventional industrial arms mainly use either suction cups or some kind of robot fingers to grasp the items.[3] Suction cups simplify the task but can't be used on everything. Developing fingers that are agile and responsive enough to handle delicate or uneven objects, such as fresh food produce, is proving challenging. Advances in soft robotics are likely to provide a solution soon.

For the moment, the unloading and loading of goods coming in and going out of the warehouses, and whizzing pallet stacks around with forklift trucks is still mainly done by humans. But it will be only a matter of time before most of that becomes roboticized too. That time will probably be sooner rather than later, following Boston Dynamic's recent unveiling of its Handle robot. Handle is a little reminiscent of a velociraptor on wheels, and is able to unload boxes from trucks, move them around, and stack them on pallets.

Another area in which rather different kinds of picking robots are being developed is agriculture. In the past few years there have been rapid developments in autonomous or semi-autonomous pieces of large-scale agricultural machinery, such as tractors, intended to spray, plant, plough, and weed on farms with vast fields. Such machines are already commercially available and in use in Australia and Canada by early adopters.[4] More compact specialized robots are also being developed for use on smaller farms with complex irregularly shaped fields. Indeed, increasing numbers are currently undergoing trials. These robots can build detailed plant maps of fields and inspect the state of crops. Being lightweight, they do minimal damage to the soil. One variety of these smaller machines is the precision weeding robot. Its visual systems can accurately identify hundreds of different weeds, and on-board lasers are used to zap the unwanted plants.[5]

There is also a range of harvesting robots in use. Robots can pick lettuces or soft fruit such as strawberries and raspberries, or even prune grape vines. Robots are at work that can identify if a piece of soft fruit is ripe and then pick it with grippers that exert just enough force to pull it off the plant without damaging it. Such machines can pick at about two to three times the rate of a human. These robots are not yet widespread, but as they become even more efficient and robust we are very likely to see increasingly automated farms, both large and small.

In September 2019 Boston Dynamics launched Spot, a robot dog that they hope will open up a new market for walking machines. Spot can climb stairs and cross rough terrain, and can be used indoors or outside.[6] It is mainly aimed at remote sensing and inspection-type applications, for instance on construction sites, or oil and gas facilities. It can also carry payloads, and it is possible to attach a six-degrees-of-freedom arm for grasping and manipulating objects. The robot autonomously locomotes and avoids obstacles using sophisticated on-board control algorithms. Currently, most applications require remote teleoperated "driving" by a user, via a joystick, but a wider menu of autonomous behaviors is becoming available.

### When will fully autonomous vehicles appear widely?

Throughout this book I have used autonomous vehicles as an exemplar case of current robotic technology but have not yet bitten the bullet on this important question. That is because there is quite a bit of uncertainty about how quickly such machines might take off. Developing autonomous vehicles that will operate safely in all driving conditions is very difficult—indeed, it may be an unrealistic goal, at least in the short to medium term. Humans do not drive well in extreme weather conditions either. A more realistic goal, being investigated by many of the major players in this area, might be to recognize when it is unsafe to continue and then to retreat to a relatively secure position. While there are already autonomous vehicles that seem to work well in restricted areas and conditions, there are still challenges in dealing with crowded, chaotic mixes of pedestrians and all kinds of other vehicles, some of which are not necessarily following the rules of the road.

The Society of Automotive Engineers (SAE) has defined six levels of autonomy (0–5) for self-driving cars.[7] This widely adopted scheme provides a useful context for discussing how far we have progressed.

**Level 0**: No driving automation at all; the driver is in control at all times. Most vehicles on the roads around the world still fall into this class.

**Level 1**: Driver assistance. The driver is always in control, but there is some limited automation through steering OR braking/acceleration support, such as lane centering, or cruise control, or parking assistance steering, but only one of steering or brake/acceleration automation can be active at a time. Many new cars now have such features.

**Level 2**: Partial autonomy. The driver is always in control and must continually monitor the environment, but some limited automation through steering AND braking/acceleration (i.e., both simultaneously) is supported, as in more advanced "auto-pilot" features or automatic parking systems.

**Level 3**: Conditional autonomy. True autonomous driving features can be engaged for prolonged periods. The car is in control, but the driver must be ready to take over at all times, as the systems are designed for very specific traffic conditions and environments (e.g., motorway/highway driving) and are not guaranteed to be able to handle all conditions. Some advanced cruise controls fall into this category.

**Level 4**: High level of autonomy. The vehicle is always capable of driving itself in certain, restricted (environmental/traffic) conditions, e.g., self-driving taxis in relatively quiet neighborhoods in normal weather conditions. A driver may still be required to make sure it is safe to enable the autonomous systems and to take over if it is not.

**Level 5**: Full autonomy. The vehicle can drive itself in all conditions, and no human driver is ever needed. No steering wheel is required.

At level 3 and above the vehicle is responsible for monitoring the environment and controlling all aspects of driving. The human driver (always needed at level 3 and as a safety net at level 4) does not have to pay attention to the road environment all the time while autonomous features are active. Level 2 vehicles are widely available and the Audi R8, which is on the streets in some countries, achieves level 3. Most other leading automotive brands are expected to release level 3 cars over the coming years. As we have already seen, level 4 vehicles, such as Waymo's driverless cars, are undergoing extensive trials in several countries, but none is yet available on the open commercial market. No truly level 5 vehicles have yet been developed: this is still an open research problem, with many teams working hard to crack it.

While level 4 vehicles exist, the conditions under which they can safely operate are probably still too restrictive for them to be widely adopted. More work needs to be done on handling adverse weather conditions: even moderate rain can confuse their sensory systems. Like so many problems in AI and robotics, level 5 is turning out to be harder than industry optimists had hoped. Accurately recognizing at all times all pertinent features in the environment, such as pedestrians, is not yet completely watertight. While modern neural network machine-learning techniques, such as deep learning, are very powerful, they are not perfect and do have limitations.

Dealing with the distorting and obscuring effects of adverse weather, or difficult, rapidly changing lighting conditions, or recognizing the difference between animate and inanimate objects, is one thing. But there are also problems arising from practices within the AI community. For instance, there are still examples of bias in training sets such that vehicle vision systems do not accurately recognize dark-skinned people as pedestrians because the examples seen during learning were predominantly white. The development of robust AI systems for decision making and planning when a tricky situation arises, such as an accident at a complex junction, or the appearance of an unexpected object in the road, is also proving challenging. AI systems do not have a good grasp of cause and effect and, as we've seen, are not able to rapidly improvise solutions based on a vast array of past experiences, generalizing from different situations (as we can). It is possible we will have to wait for the next revolution in machine learning before some of these issues can be solved. On top of these technical demands is the big question of legal frameworks under which autonomous vehicles must operate. These may well vary from country to country, at least initially until the technology is fully mature. And then people will have to decide that they trust the technology with their lives, if it is going to be widely adopted.

Taking all this into account, I think it is almost inevitable that level 4 autonomous vehicles will appear quite widely, quite soon—by which I mean in maybe a decade, maybe two— but their operating conditions may well be far more restrictive than some had hoped. True driverless level 5 autonomy—that is, being able to drive safely on any road that a human can manage, in all but the most extreme conditions—could be quite some way off yet, possibly several decades at least. But level 5 within restricted zones may well come sooner than that.

The arguments for autonomous vehicles revolve around safety (most accidents are caused by human error), environmental impact (all such vehicles will be electric), reduction in congestion (less need for parking in city centers, more efficient driving), and convenience. One way to aid all these things is to simplify the autonomous driving problem by changing the environment. So, echoing aspects of the early-twentieth-century schemes briefly discussed in Chapter 4, there might be well-defined geographical zones with extra infrastructure (possibly laid in the roads, possibly at the road side) that increase safety and efficiency while simplifying the driving problem. Level 4 (and eventually level 5) vehicles might only be able to operate in such zones. An element of external control, through information from sensors that will be part of the extra infrastructure, could be included. Or at least data on traffic density in nearby locations might be fed to all vehicles and taken into account by control and planning algorithms aimed at maximizing flow and eradicating gridlocks, perhaps building on ideas from decentralized control in groups of robots.

Rodney Brooks, influential roboticist, robot entrepreneur, and general AI guru, has made some interesting predictions about when autonomous vehicles will start to take off. In contrast to what might be politely described as over-enthusiastic claims from some working in the field (mainly in marketing), Brooks' prophesies are decidedly level-headed. He predicts that a proper driverless taxi service—i.e., one with arbitrary pick-up and drop-off locations—will appear in a major US city in the next couple of decades, but no earlier than 2032, and even then probably only within a geographically restricted area. More boldly, he predicts that by 2031 "a major city will ban parking and cars with drivers from a non-trivial portion of the city so that driverless cars have free reign in that area."[8] He believes that "this will be the starting point for a turning of the tide towards driverless cars," and that in time the majority of US city centers will operate under such rules, but not before 2045.

It is true that for a few years several companies in the USA have either been trialing driverless taxi services or planning to do so in the near future. However, many of these are not really taxi services and not fully driverless. Some have designated, marked stops for boarding, so are essentially bus services, and nearly all feature on-board human drivers for safety. However, in 2019 Waymo substantially moved things forward from its earlier limited taxis by starting to operate a true taxi service in a suburb of Phoenix, Arizona, with no human drivers on-board.[9] This service is totting up about 1,500 monthly users, but it is not yet open to the public. Users must be pre-approved and have to sign non-disclosure agreements barring them from talking about any mishaps during a ride. All vehicles in the service operate with a remote safety driver monitoring the journey, and they do not run when it is raining. Thus this is an advanced level 4 autonomous taxi service that works in a relatively quiet and restricted area as long as it is not raining. That is impressive but not yet quite at the game-changing level, the weather limitation being particularly problematic.

Several manufacturers are developing autonomous trucks. Trials (with safety drivers on-board) have started on stretches of road in Texas.[10] There is talk of dedicated lanes in which convoys of truly self-driving trucks will cruise. Regulations, safety testing, and the technical limitations of current AI all point to the likelihood that this will not come to pass overnight. Controlling long-distance highway driving has fewer technical challenges than dealing with busy urban conditions, but the damage that could potentially be done by a malfunctioning autonomous eighteen-wheeler is colossal. However, advanced level 3 or level 4 trucks, with drivers in place prepared to take over, will probably appear quite soon, although there may be restrictions on the conditions and areas in which they are allowed to operate.

As briefly mentioned in Chapter 1, there have been recent experiments with small driverless vehicles intended for so-called "last mile deliveries." The vehicles operate on routes within small, quiet areas—at the moment mainly on footpaths rather than roads. They stop outside a residence, the customer unloads their goods, and the vehicle continues on its route or returns to base. Such vehicles have the potential to become a nuisance if they clutter up pavements (where they have to move slowly) or block roads. But in certain quiet neighborhoods with wide roads and paths they might become popular.

So, autonomous, semi-autonomous, and close-to-autonomous vehicles are set to have a significant impact on road transport and our relationship with it. Private ownership of cars may well decline as a result, as will congestion and environmental impact. But this will all take time. Be skeptical of claims that it will be with us in a year or two.

### What about robots in health and social care?

Robots are beginning to invade medical spaces. Autonomous robot carts are appearing in some hospitals. They trundle around corridors carrying supplies from one place to another, either following navigational markers on the floor and walls or using LIDAR maps. They are expensive, so are not yet widely used. Remotely operated telepresence robots that allow doctors to examine patients in far-away areas with limited access to medical facilities are becoming more common. Robotic surgical tools that can enhance a surgeon's precision and enable work at ever-shrinking scales are becoming widely available in developed countries. These are surgeon-operated robots, but there is research into making them more autonomous, which will open up some meaty legal and ethical challenges. Robotic surgery is an area where many commercial companies and university research groups are fast developing new techniques and innovative technology, so it is safe to assume it will continue to grow. As part of the fight against antibiotic-resistant bacteria, some hospitals in the USA are starting to employ disinfecting robots that move autonomously to recently vacated rooms, following a patient discharge, and then bathe the area with high-power pulsed ultra-violet rays to obliterate all microorganisms.[11]

The problem of an aging population has long been a driver in Japan for the development of robots that can help with old-age care. In some care homes, residents pet and cuddle robotic animals such as the robot dog AIBO and Paro, a furry robot seal.[12] This can provide therapeutic support to those suffering from dementia, by soothing anxiety and keeping them engaged. Humanoid robots, such as Pepper, give pre-programmed exercise classes, as well as engaging in scripted conversations. In some homes human carers wear powered exoskeletons to help while lifting patients.

Such approaches are increasingly being investigated elsewhere in the world, particularly for dementia care, where they can also be used to give some relief to home carers. The focus is on robots that can assist human carers rather than replace them.

Healthcare is a potentially enormous market, so, for commercial reasons alone, this is an area that is likely to see a large increase in the use of robots for cleaning and routine tasks, and more advanced robotic tools and robot assistants to aid and augment medical and care professions, and to ease the burden for carers.

### How are robots used in education?

By far the most popular use of robots in education—both at school and university level—is as a tool to teach aspects of science, technology, engineering, and mathematics (STEM) subjects. Mobile robots, usually simple wheeled varieties, can be constructed and/or programmed to act in the world. Elements of computer programming, engineering, and mathematics can thus be learnt in an engaging, fun project.

For the past decade or so, there have been experiments with using humanoid robots as a therapeutic aid for children with some kinds of autism.[13] Such children can become overwhelmed by social interactions, including with their teachers and therapists. This can make it difficult for them to focus and learn both academic and social skills. Often they appear to find it easier to interact with robots and technology. Some researchers have been experimenting with socially responsive humanoid robots as a way of teaching the children how to more comfortably interact with their fellow humans. Others have used such robots as a way of teaching academic material to children with special needs. While some researchers are cautiously optimistic, there are disagreements about how effective such therapy is in the long run, and whether increased engagement comes mainly from the novelty of using a robot.[14]

For many years, various kinds of computer-based tutoring systems have been used to help deliver lessons in the classroom. More recently, researchers have been investigating the use of social humanoid robots to deliver mainstream education, behaving as tutors who interact with the pupils.[15]

On restricted tasks they have achieved outcomes similar to those of human tutors. There is evidence that this is largely due to their physical presence: their embodied nature makes them more engaging, more exciting, than standard computer-based tools. So far, success has been achieved in fairly narrow domains, and these robots would be unlikely to fare well in an unsupervised class of unruly 13-year-olds on a Friday afternoon. But some commercial products have resulted, such as robots for teaching basic English. It is not yet clear if the effectiveness of such robots declines as the novelty wears off.

### How are robots used in the military?

Robots are increasingly deployed by the military. The first serious uses of such technology date back at least as far as the Second World War, when the Russian army began employing radio-operated unmanned tanks. Remote-controlled military robots, such as the rugged track-driven PackBot,[16] with multiple sensors and an articulated arm and gripper, are now commonplace. They are used in search and rescue missions, to handle dangerous materials—such as in bomb defusing, to transport equipment, and for surveillance. Modern unmanned aerial vehicles (UAVs) and larger remotely piloted aircraft (RPAs) are remote-controlled flying robots; they are widely used by most military forces. Some carry lethal payloads and are used for attack as well as reconnaissance.

Armed autonomous mobile robots and wider autonomous weapons systems are being experimented with by various armies. Some advanced missile systems are essentially deadly autonomous flying robots. A sizeable number of commanders and politicians see the future of military conflict as battles between robots, some autonomous. There are huge dangers and deeply worrying moral and ethical issues if that road is taken. These issues are discussed further in Chapter 8.

## Can robots be used in art?

Although the profit motive is the ultimate driver in many robot developments, it is important to realize that there are other more cultural uses too. As we have seen, from their origins in literature, they have become a mainstay in films but are also increasingly being used in works of visual art, or even as creators of art. Such works question our fascination with self and other, and probe the meaning of authorship.

A powerful early example was Edward Ihnatowicz's 1970 work *The Senster*. Described by Ihnatowicz as a work of cybernetic sculpture, *The Senster* was a large (15-ft-long) articulated skeletal structure built from tubular steel. It loosely resembled a tall spindly crane-like bird, or perhaps a skinny upright reptile.[17] Controlled by a digital main-frame computer and powered by hydraulics, the robot had four microphones and two radar sensors mounted in its head. It could thus respond to sound and movement. The microphones could be steered independently of the head and would home in on the dominant sound in the environment, the rest of the structure following in stages if the sound persisted. The robot would track moving noises (maybe laughing children running around the hall, or a pair of high heels click-clacking across the floor), but loud sounds and sudden movement would make it rear up as it shied away. Built by Ihnatowicz and collaborators at the Department of Mechanical Engineering, University College London, the piece was commissioned by the electronics giant Philips, to be installed in its Evoluon science center in Eindhoven, the Netherlands. It was on permanent display from 1970 to 1974. Although its underlying behaviors were simple, the combination of the fluid, eerily life-like movements of the robot, the unpredictable behavior of the audience, and the intricate acoustics of the hall resulted in what appeared to be complex interactions between *The Senster* and the crowd of viewers. This made it a very popular exhibit.

Since the pioneering efforts of Ihnatowicz, and a handful of other cybernetic artists of the 1960s, there have been many examples of interactive, kinetic robotic art. These include the thought-provoking works of Simon Penny and Ken Rinaldo, the large-scale subversive spectacles of Survival Research Laboratories, sometimes complete with explosions and fire-breathing machines, the provocative, sensorially rich explorations of Laura Dekker, and the recent *Senster*-like experiments with Alter-3, a humanoid robot developed in 2019 by Hiroshi Ishiguro, Itsuki Doi, and Takashi Ikegami (Figure 7.1). Alter-3 is autonomous, using artificial neural networks, and responds to its environment. Although the AI and robotic engineering are in many ways more advanced than in *The Senster*, and Alter-3 is an impressive machine, to me the movements of the older robot seem more captivating and life-like.[18]

**Figure 7.1** Alter-3.

*Source:* © Hiroshi Ishiguro, Takashi Ikegami, and Itsuki Doi. Tristan Fewings/Getty Images for Barbican Centre.

Over the years there have been a number of attempts to develop drawing robots. These range from mobile robots with attached pens making marks as they move across a surface,[19] to robot arms, or even humanoid robots, grasping pens and making more "conventional" drawings. They operate according to some set of rules or more or less complex algorithms, possibly under their own autonomous control.

Good examples of arm-based drawing machines are Patrick Tresset's series of robots—articulated arms located on a desk that hold pens and sketch on paper.[20] Moveable cameras take images of the scene to be drawn, perhaps a person's face for a portrait, and can "observe" the progress of the arm. The camera images are analyzed and the robots draw according to algorithms developed over many years by Tresset. These machines form part of a lineage that stretches back to computer art pioneer Harold Cohen and beyond. Starting in 1973, Cohen worked over several decades on a drawing program, named AARON, continually refining it. It produced engaging, mainly abstract, works that were physically produced on large flatbed plotters, controlled by Cohen's algorithms.[21]

Such drawing machines raise interesting questions about creativity and whether the robot or the developer of the robot control system, or neither, is the author of the resulting artwork.[22] In the case of robots controlled by fixed, non-adaptive algorithms designed by the system's developer, the robot might be viewed as an output device, the creator of the art being the author of the algorithms. But in the case of robots controlled by systems produced by non-deterministic adaptive processes—such as evolutionary or machine-learning methods—where the possible outcomes of the overall process cannot be predicted and will be distinctly opaque to the system designer, then who or what is the creator of the art is less clear. Because the control system has been developed by an adaptive mechanism, responding to unpredictable environmental forces, then the resulting drawing might be considered as the output of a complex emergent process, rather than a deliberative cognitive act.

Robots have also spread into other areas of art and performance. In the mid-2010s a number of composers, including Evelyn Ficarra, developed operas to be performed by robots. Ficarra's work, *O, One*, for two humanoid Nao robots and cello, and partly sung in binary code, arose from her interest in how opera can survive and respond to new technology.

### How are robots used in pure science?

Robots have long being used for purely scientific reasons—as tools for planetary exploration, and the study of other hostile environments; and as precision aids in exquisitely delicate experimentation, for instance in the exact positioning of an array of chemical droppers, as a way of handling dangerous materials. Since Grey Walter's pioneering work with his tortoises, they have also been used in the synthetic approach to biology, that is, as physical models of biological creatures, to allow the investigation of whole behaving systems, to illuminate mechanisms that would be too difficult to probe in living animals[23] These include the neural processes underlying certain behaviors, the strategies used by animals to complete various tasks, and the ways in which communication within a group of organisms can operate.

A very nice recent example of such work is the development of a "PigeonBot" by scientists at Stanford University. Their primary aim was to understand more about the mechanisms of flapping-wing flight in birds, in particular how the wing surfaces dynamically change shape during flight. The team developed a brilliantly conceived biomechanical hybrid robot—the PigeonBot—by incorporating real pigeon feathers into the wing of the flying machine.[24] This idea paid dividends. The study revealed a lot about how birds fly, particularly how feathers and the ways they are connected enable the powerful mechanisms of continuous wing morphing. It also points the way to new forms of bio-mimetic flying robots that could have a lot of useful applications. Wing morphing considerably improves efficiency and maneuverability in nature, so further developments of such robots might lead to a new class of agile flying robots that are superior to standard drones.

Another striking example is the work by Lee Cronin's team at Glasgow University on using a robotic system in the automated discovery of chemical reactions.[25] The discovery of new chemical reactions and potentially useful new chemicals is a very time-consuming business. Trying out numerous concoctions until something interesting appears involves a lot of random trial and error. Experienced chemists have hard-won knowledge and intuition to guide the process, but it can still be a long slog. The Glasgow group devised a robot to automate the procedure. The key idea is that the robot learns about the reactivity of combinations of various organic chemicals by doing its own experiments. The system is then able to make predictions about how reactive other combinations might be. That is, it automatically builds up a kind of picture of the reactivity of the chemical space. This enabled human chemists to focus on a small number of candidate reactions the robot predicted would be interesting. In this way, Cronin's team was able to discover four previously unknown reactions.

The robot controlled a set of pumps and valves which allowed it to combine liquid chemicals, from an array of reservoirs, in various combinations in reactor flasks. It was able to do six such experiments in parallel. The chemical reagents and the resulting mixture (which may or may not have produced a reaction) were automatically analyzed using infrared and nuclear magnetic resonance spectroscopy units that were incorporated into the system. A machine-learning system took these various spectra as inputs and learnt to predict whether or not a given mixture would be reactive. As the system learnt and started to map out the chemical space, it was able to make more and more informed decisions about which experiment to try next. After performing about one hundred experiments the robot system was able to predict the reactivity of a set of one thousand possible combinations of reagents with an accuracy greater than 80%.

The robot explored the chemical space much more efficiently than a human could. This work points the way to similar robot chemistry applications where the search of chemical space could be automatically focused into areas that promise the discovery of new materials with particular desirable properties.

An exciting application of autonomous robotics that is starting to come into its own is the exploration of outer space. A series of NASA Mars Rovers, which became gradually more autonomous with each subsequent mission, have been used to collect and analyze samples and conduct simple experiments since the late 1990s. *Curiosity*, which landed on Mars in 2012, is still actively exploring the red planet's geology and climate.[26] Because of the significant delays and complications in sending and receiving signals between Earth and Mars, it is much more practical to use autonomous robots that can be periodically guided from mission control. *Curiosity* uses autonomous navigation and obstacle avoidance to explore Mars within parameters set by NASA engineers on Earth, and can control its array of sophisticated scientific instruments. In February 2021 a new Mars mission successfully landed on the planet. The robot rover *Perseverance*, whose design is derived from that of *Curiosity*, has been deployed. This mission includes an autonomous flying helicopter robot, *Ingenuity*, which is to be used to explore the practicalities of powered flight on other planets. As missions go further afield and close contact becomes impossible, autonomy and more sophisticated AI will be crucial.

## Will robots take over our jobs?

Robots have been replacing humans in various jobs for decades, mainly in heavy industry. Industrial robot arms are faster, more accurate, more efficient, and ultimately cheaper than the human labor they have taken over from in some sectors. They can operate in conditions that would be intolerable to us. Such automation increases productivity and enables more complex industrial processes. But, so far, this practice has tended to have a net effect—expanding industries and creating new jobs within the same organization, or in the enlarged supply chain, or surrounding service industries, as well as creating new markets and jobs in robot technology. However, the replacement job is not always at the same skill or pay level or in the same location. As new kinds of robots appear in sectors beyond manufacturing this model may not hold, and the amount of available work might decrease.

There have been a number of recent studies, widely reported in the media, that have tried to predict where jobs will disappear. Many of the media reports have used rather misleading language, referring to jobs being taken over by robots when the studies were actually about the more general phenomenon of automation. Automation includes the use of physical robots, but much of the focus was actually on replacement by other technologies, including disembodied AI computer systems.

Depending on the assumptions made and data used, headline estimates from these studies vary from about 9 to 47% of jobs in the USA at high risk of automation, and 7 to 38% in the UK. How were those numbers arrived at? Are they reasonable? How is it even possible to make those kinds of predictions? It is worth spending a little time looking behind the headlines. The aim in each case was to build a prediction model: that is, to provide a fairly complex equation into which the values of a set of variables describing a job are fed. Out pops the predicted risk of that job being automated in the fairly near future (say the next 10–20 years). At the root of this process is the set of variables representing the pertinent features of a job. In 2013 Carl Frey and Michael Osborne, from the University of Oxford, did the first major study of this kind.[27] They chose nine variables that they felt best described the dimensions of a job that made it potentially hard to automate (high values would mean low risk of automation and vice versa), and that were included in large databases of job descriptions maintained for national and international labor market analysis.[28] The variables related to perception and manipulation (such as finger dexterity), creative intelligence (such as originality), and social intelligence (such as empathy and persuasiveness).

A group of AI and robotics experts gave "susceptible" (to automation) and "not susceptible" labels to a subset of 70 occupations from a database of more than 700. These labels (basically informed guesses), along with the values of Frey and Osborne's chosen variables, taken from the database, were used to train a statistical model-building algorithm (essentially a form of machine learning).[29] The database variable values were derived from responses to surveys of various economic sectors.

The model thus produced was run on the other jobs in the database to give the predictions. Later studies used similar methodologies based on different databases and different assumptions relating to ways of breaking jobs down into constituent tasks and taking into account subcategories of jobs within the same occupation.[30]

Although this is state-of-the-art model building, like most exercises in forecasting it is noisy, murky, and riddled with assumptions. Trying to compress complex socio-eco-political forces into a relatively simple mathematical equation is never wholly satisfactory. There are multiple points at which subjective judgments are made (the expert labels, the values returned by survey respondents, the model variables chosen, and so on). It is no surprise that the estimates vary so widely.

Some of the jobs identified as being at high risk of automation, such as retail assistants, are flagged up less because robots are about to start serving you in the local mall, but more because retail outlets (particularly supermarkets) are already eradicating or simplifying jobs by offloading much of the work onto customers (e.g., the self-service checkout). This trend is spreading to some kinds of restaurants; at others we see the beginning of mobile robot waiters taking orders.

The more recent studies[31] agree that the 2020s will see an acceleration of robots taking over routine tasks in structured, controlled environments, such as moving objects around warehouses. They predict that as the 2020s draw on and into the 2030s, we will probably see a wave of robotic automation of other kinds of physical labor, including tasks that require manual dexterity, in sectors such as construction. Autonomous or semi-autonomous robots, capable of some degree of problem solving in dynamic, noisy real-world situations, will become more prevalent in transport as well as other areas.

In some sectors, including healthcare and education, we might expect to see robots becoming used as tools to aid and augment, rather than replace, humans, to transform jobs rather than destroy them.

### Should we let them take our jobs?

This is a very complex question, and it is not new. At various times during the twentieth century there were moments of anxiety and public debate about the consequences of automation. During the 1920s and 1930s, as a wave of production automation arrived in some industries, and then again in the 1950s as the world recovered from the Second World War, there was an assumption that increased productivity due to automation would lead to shorter working hours and a better life for many. During this period the five-day working week became common, partly as a way of maintaining employment levels in the face of automation. This led to a better balance between leisure and toil, although there had been patronizing discussions among the ruling classes about whether or not the working classes could be trusted to make sensible use of any extra free time. There was a hope among many that increased automation and efficiency would lead to the end of scarcity, to a reasonable minimum standard of living for all. At that time, many large employers (but by no means all) had a fairly benevolent attitude toward their workers, taking their social responsibilities seriously. As we enter a new wave of robotic automation, social and economic structures have shifted. Globalization and increases in the scale of corporations mean that many employers are now more detached from their workers, particularly the low-skilled ones. There is often pressure to reduce costs and increasing profits, share price, and dividends, resulting in less-caring employment practices. The abdication of social responsibility, through complex schemes to minimize tax liabilities, is widespread among the tech giants and other new industries.

There are three broad possible directions of travel that are often talked about. The first is to carry on regardless and assume, like the first wave of robotic automation in heavy industry, that the current one will create at least as many jobs as it destroys. The second, utopian, vision is to grasp the opportunity to create much more leisure time and abolish scarcity, to provide a good standard of life for all. The third trajectory involves advanced automation increasing the wealth of certain sections of the population—the rich, the skilled, and the highly educated—while the rest become poorer. This trajectory might be the inevitable result of a lack of planning if the first path does not come to pass. It would turbocharge inequality.

It is very likely that as the next generation of robots, including autonomous vehicles, become embedded in various industries, less-well-educated and lower-skilled workers will be particularly vulnerable to job loss. This highlights the importance of both companies and governments investing in lifelong learning and retraining.[32] As robots, and AI technology in general, slowly become more intelligent, we will see them used more and more to augment various skilled human roles, and it will get to the stage that they could take over certain positions that are now seen as skilled, for instance in finance, law, and management.

There are a growing number of advocates of various systems of universal basic income whereby increased automation could be used to create wealth that is then distributed so as to avoid massive inequality.[33] Everyone is provided with a basic income (possibly with some strings attached) that allows a reasonable standard of living. People can either spend more time on leisure and creative activities or education and retraining, as many jobs become part-time, or are free to earn more money as they pursue their career or business goals. Current economic and political structures and attitudes in some countries, such as the USA, would probably make such a system very difficult to introduce; it would be deeply unpalatable to many with power, influence, and wealth.[34]

In many of these scenarios, huge alterations to society will have to be managed, shaped, and regulated. In most countries that means using the political sphere to take democratic control of the forces at play. The alternative is to sleepwalk into a future fashioned by and for huge global corporations. As with the adoption of any powerful technology, there will be consequences—some intended, some not.

### Will robots fundamentally change the way we live?

Yes, I think they will. They will gradually transform the nature of many jobs—either taking them over or changing their form by being used to augment human abilities. They will slowly become commonplace in homes and on our roads and streets, and will become tools for care and education. How long this transformation will take is still very uncertain, but make no mistake: it has started.

The widespread use of robots has the potential to open up more leisure time and opportunities for many people. It also has the potential to create a dystopian world of super-elites and underclasses. Depending on how we respond to the various socio-political challenges, inequality could either escalate or be eradicated.

# 8

# ROBOT ETHICS

## What is robot ethics?

Broadly speaking, ethics is the rational analysis, shaping, and application of moral principles. It is formally studied as a branch of philosophy, but many believe that the ability to think critically about moral values is an inherent human ability. Concepts of right and wrong behavior and how to use them to guide our actions are at the heart of ethics.[1] Robot ethics, therefore, is the application of ethics within robotics, guiding how we design, use, and interact with robots, and how they behave toward us. It encompasses both our behavior, in the ways we apply robotics within society, and the behavior of robots: how can we make a robot act in an ethical way? In the future, could a robot bear moral responsibility for its actions? It is an inherently multidisciplinary endeavor involving contributions from engineering, philosophy, law, psychology, and sociology, among others.[2]

Robot ethics is a sprawling, tangled field, full of unresolved complexities. In this chapter I can only touch on a few of the main issues, giving a rough introduction to current concerns.

## Why is robot ethics important?

Twenty years ago most of us who worked in commercial or academic robotics and AI hardly gave robot ethics a second thought. There were intellectually interesting debates to be had, for sure, but they seemed rather abstract and disconnected from the real world, because the new generation of robots, particularly autonomous, primitively intelligent robots, had hardly ventured out of the labs. But that has all changed. Now that such robots are starting to appear on our roads and in our homes, and are being used by the military, big ethical, legal, and philosophical questions are suddenly very pertinent. A pedestrian has already been killed by a car operating in autonomous mode.[3] There are major concerns about potential military uses of autonomous robots. Once robots are widespread and commonly share our environments, even barely intelligent machines can cause serious legal and ethical issues—as can misplaced faith in over-hyped, under-regulated technology. Many of these topics have been explored in literature and films for decades; now it is time to deal with them in reality.

Over the past few years, all over the world various governments, research funders, academic institutions, and other bodies have set up initiatives, boards, networks, think-tanks, and talking shops to grapple with the ethical problems arising from robotics and AI.[4] What was once a backwater is now a very hot topic.

### Should robots be socially and morally responsible?

Unquestionably. Robots intended to integrate into our everyday and work lives must be designed to follow the moral codes accepted by society, the ethical norms of the day. In the case of pre-programmed, fixed, dumb machines, such as industrial robot arms, this is relatively straightforward. They must be designed to be safe. No one in their vicinity should be put at risk by the operations of the robot. They should not emit dangerous chemicals in the presence of humans, and their movements must not endanger nearby people, and so on. Things can get a whole lot more complicated when the robots become mobile and autonomous and make their own decisions.

As briefly mentioned earlier in this book (see Chapter 5), one of the first attempts to provide ethical rules for intelligent robots was Isaac Asimov's Three Laws of Robotics, which featured in his 1942 story *Runaround*:[5]

1) A robot may not injure a human being or, through inaction, allow a human being to come to harm.
2) A robot must obey the orders given it by human beings except where such orders would conflict with the First Law.
3) A robot must protect its own existence as long as such protection does not conflict with the First or Second Laws.

On the face of it Asimov's rules seem sensible. But it doesn't take long to see that there are many situations that would set up unresolvable conflicts. This made them very useful as a plot device in Asimov's fiction, creating dilemmas and unintended consequences that generated jeopardy. Apart from being inadequate, these very general laws assume highly sophisticated intelligence on the part of the robots, which would have to make complex judgments. The story was set in 2015, but we are still very far from being able to create the required level of robot cognition.

The inadequacy of the first law is exposed by the following very real dilemma that might be faced by a fully autonomous vehicle. Imagine an autonomous taxi carrying four passengers traveling along a road at the top permissible speed. Suddenly a child runs out in front of it. The taxi's AI systems instantly calculate that the braking distance is too long to avoid hitting the child if the vehicle continues straight ahead. Its planning systems reason that if it swerves one way it will shoot into the path of an oncoming heavy goods truck, putting the lives of all passengers at severe risk, along with that of the truck driver. If it swerves the other way it will plough into a crowd of pedestrians and then hit a solid brick wall, almost certainly causing multiple deaths. In the face of this horrible dilemma what should the AI system do? What would a human driver do, with a split second to react? And would there be a different moral choice if instead of a child running into the road it had been a dog?

In the case of a human driver, where there can be no good outcome in this scenario, the whole thing will be put down to a tragic accident whatever they do. But in the case of an autonomous vehicle, which generally has the advantage of being able to react faster than a human, there might be an expectation that the AI control systems did the best they could according to some acceptable moral criteria. Many have argued that making robots moral should involve codifying certain ethical principles into rules of behavior that the machine must follow in certain circumstances. These might involve adopting a particular theory of ethics such as utilitarianism, which is based on maximizing some measure of the consequences of action, essentially favoring outcomes that produce the greatest good to the greatest number.[6] Or the behavioral rules might include the doctrine of double effect, an ethical principle that has long been (at least partially) enshrined in secular and religious law. The doctrine of double effect states that if doing something morally good (e.g., saving many lives) has a morally bad side effect (the double effect, e.g., killing a smaller number in the process) it can still be ethical as long as the bad side effect was not intended, even if its likelihood was foreseen. Common examples are unintended harm arising from self-defense, or the use of drugs in palliative care to ease severe pain even though such treatment might speed the inevitable end; or indeed trying to minimize the number of deaths in a car crash scenario.

Programming-in such general principles that the robot would be expected to apply in the correct way in a wide range of circumstances is beyond current AI technology, as it would include complex predictions and judgments. This is one of the important challenges that arise when trying to develop more general AI. The best that could be done at the moment would be to design specific versions of the rules for very specific circumstances that the robot might be expected to encounter in the narrow role it is designed to fulfill.

Designing specific rules of ethical conduct is itself not necessarily straightforward. The theories of what philosophers call normative ethics—that is, the practical ways of deciding what is a moral course of action, which include utilitarianism—are often incomplete and convoluted and defer to vague concepts such as common sense; in other words, they are unsuited to codifying as clear rules. Even worse, there are a number of different, sometimes conflicting, theories. As a society we often implicitly employ several theories simultaneously in the ways we deal with risk and safety.[7] Some have argued that developing systems of rules governing the ethics used by autonomous robots would do much to advance ethics in general as it will force a greater degree of precision and rigor and fill some of the gaps.[8] Others contend that while the messy organic way in which ethical norms gradually develop renders them inappropriate for boiling down into a set of rules, they may be suitable for the application of machine learning. The idea is for a robot's AI systems to learn ethical behavior by observing what we do (or are supposed to do), finding patterns in numerous training examples. While in theory this might work, it opens up fresh ethical concerns. The training sets must be unbiased, containing suitable examples of ethical and unethical behavior, and be sufficiently wide to capture all relevant behaviors. If such an approach could be made to work, how will the decisions made by the robot be rendered transparent, that is, easily explainable? Although it is difficult to capture moral codes as sets of logical rules, it is at least as difficult to penetrate the reasoning of the kinds of large artificial neural network used in many machine-learning systems. If the robot's ethical decision making is not explainable, can we trust it?

All of this begs the question, who should be responsible for deciding what system of ethics is built into, or learnt by, a robot? The robotics company CEO? Some hapless junior engineer charged with implementing an "ethics module?" Politicians and policymakers? How can we be sure that the ethics will be acceptable to society at large? Existing legal frameworks will be less help here than might be hoped. The law on ethical questions, such as those relating to the intention behind an action, is quite often foggy, confused, and incomplete, and can vary considerably from country to country.

If one autonomous car on the market uses a system of ethics that prioritizes the reduction of risk to its passengers above all other considerations, and another is programmed to minimize overall damage and loss of life to all, whether or not they are in the car, which would you buy? Should you have the choice, or should there be an agreed standard that applies to all autonomous vehicles? If, in trying to minimize overall risk to life when an accident unfolds dead ahead, an autonomous vehicle preferentially swerves toward a large new car on the left, rather than an old small car on the right (or vice versa), because that is what its ethics rules tell it to do, will the general public think that is fair? Hardwired into its behavior is a (premeditated) rule that says aim for one kind of vehicle rather than another. Would a more random response be fairer? These issues are not at all easy to resolve. Input from a range of voices to establish enforceable national and international standards seems appropriate, rather than just allowing automotive and robot companies to do as they wish. It will be important to make sure society understands what ethical standards are being applied by keeping them transparent.[9]

However the system of ethics used by a robot is decided on, and then instilled in the machine, its specifics will likely be task dependent and will reflect the role the robot has been designed for. For instance, future robots involved in care of the elderly or sick might only be very rarely involved in life or death decisions, but they would have to be sensitive to a patient's emotional state. It would be unethical to forcefully handle an elderly person with dementia if they were being uncooperative due to extreme agitation. More subtle calming techniques would have to be first employed before attempting to move them. The robot would have to be able to distinguish distress from mild grumpiness, and act appropriately.

Because we will have to try to turn moral principles into rigorous, unambiguous algorithms, in developing robot ethics we might well end up holding the machines to higher standards than we apply to ourselves. This would probably have the useful side effect of increasing acceptance of the technology, and maybe even improving human behavior.

### Should roboticists be socially and morally responsible?

The other side of the robot ethics coin is of course the people who design and build them. Robots will only behave ethically if people make them that way. At the moment, we are not capable of making robots that are ethical in the sense of having sophisticated ethical agency—the ability to make complex moral judgments. Perhaps one day we will be able to create such machines, but for now all we can do is design robots to behave ethically in their narrow range of behaviors, to give the illusion that they are ethical in the deeper sense. Hence, it is incumbent on roboticists to engage with the ethics issues and act responsibly in both robot development and the uses of robotics they help enable. In his 1985 presidential address to the Society for the History of Technology, Melvin Kranzberg discussed a set of truisms that had underpinned his work for decades and had become known as Kranzberg's laws. The first, and most famous, law runs as follows: Technology is neither good nor bad; nor is it neutral.[10] By that he meant that technology always interacts with the wider human social ecology, such that it frequently has consequences that go far beyond its immediate purposes, and it can have quite different outcomes depending on the contexts and circumstances under which it is used. Technology is often mired in unintended consequences and unforeseen complications. This is undoubtedly true of robotics, particularly AI-based autonomous robotics. Problems with bias in machine-learning training sets remind us that AI is only as impartial as its creators. When producing new autonomous robotic technology for use in society, poorly thought-through robot ethics—or worse, completely ignoring the ethical dimensions altogether—is a potential recipe for disaster.

In 2010 a small interdisciplinary workshop was held in the UK, sponsored by government research councils, at which a group of academics spanning a range of disciplines (plus one industrialist) spent a couple of days trying to develop a set of principles for designers, builders, and users of robots. The intention was to produce a starting point to spark debate and later improvements. The result was the following set of principles,[11] which emphasize the responsible role roboticists should take:

1) Robots should not be designed as weapons, except for national security reasons.
2) Robots should be designed and operated to comply with existing law, including privacy.
3) Robots are products: as with other products, they should be designed to be safe and secure.
4) Robots are manufactured artefacts: the illusion of emotions and intent should not be used to exploit vulnerable users.
5) It should be possible to find out who is responsible for any robot.

In 2016 these principles were incorporated into the first-ever explicit ethical standard for robotics, published by the British Standards Institution as BS8611: *A Guide to the Ethical Design and Application of Robots and Robotic Systems.* This standard offers guidance on how designers can undertake an ethical risk assessment of their robotic system, and how they might mitigate any issues identified. It highlights a number of ethical hazards which it groups under four broad categories: societal, application, commercial and financial, and environmental. Under the societal heading, issues such as loss of trust, deception, infringements of privacy, addiction, and loss of employment are all explicitly mentioned as things ethical roboticists should consider and guard against.

At about the same time BS8611 was published, the Institute of Electrical and Electronic Engineers (IEEE) launched its own, global, initiative on the ethics of autonomous and intelligent systems. This led to a substantial document, *Ethically aligned design: a vision for prioritizing human well-being with autonomous and intelligent systems*,[12] which put humans firmly at the center of the picture. As Alan Winfield has pointed out,[13] this is a bold and explicitly political stance, as it views robotics and AI as technologies primarily for the greater good of humanity, rather than just for cost cutting and economic growth. This document, its updates, and a number of new standards that IEEE has drafted, emphasize the importance of the ethical design and ethical use of autonomous systems.

In the numbered list given above, the first principle worries about killer robots but does not entirely prohibit them. Remotely controlled robots are becoming widespread in conflict; the responsibility for target selection and other acts with profound moral undercurrents is firmly with the human operator. But in the case of autonomous robots armed with lethal weapons, on-board AI is responsible for all decisions. AI systems are not perfect, often far from it. Even in normal calm conditions, can AI be trusted to always tell the difference between civilians and combatants? Not without significant risk. In the complex, noisy, dynamic, unstructured environment of battle, with dust, smoke, and fire distorting and obscuring sensors, would autonomous robots be capable of accurate recognition, let alone making the complex ethical choices that arise? Many of us believe they would not, that developing autonomous weapon systems is irresponsible and crosses a moral line. Completely replacing troops with machines, which is the vision of some generals, might lower thresholds and make the decision to go to war easier, consequently further shifting the burden of conflict on to civilians, who will inevitably be caught in the chaotic currents that will be unleashed.

Such concerns led Noel Sharkey, Jurgen Altmann, and Peter Asaro to found the International Committee for Robot Arms Control (ICRAC), which argues, among other things, for the "prohibition of the development, deployment and use of armed autonomous unmanned systems; machines should not be allowed to make the decision to kill people."[14] ICRAC is part of the wider Campaign to Stop Killer Robots which lobbies against the use of autonomous weapons. The Campaign has been successful in initiating discussions of autonomous weapons at the United Nations Convention on Conventional Weapons, and is pushing to advance those talks to treaty negotiations.

A further worry is that if autonomous armed robots become normalized in the military, they will start to seep into the police, then private security firms. What if they get into the hands of criminals or terrorists?

### What about cases where ethical robots are used in an unethical way?

It is not hard to imagine various situations where a robot is used outside its intended purpose, resulting in unethical behavior on the part of the user. For instance, suppose a robot tutor was designed to give children occasional short lessons within an overall educational framework where humans do most of the teaching and are in charge of supervision. If the robot was instead used as the sole source of education and supervision for a group of children, that would amount to unethical behavior by the parents or whoever put the children in the robot's care. In some circumstances, legislation might have to be tightened up to prohibit such things. Even if future robot tutors and carers became much more sophisticated and could responsibly and safely teach and look after children or the unwell or infirm, there is the question of whether or not over time this would have a dehumanizing effect on those being taught or cared for. If children end up spending most of their time with humanoid robots, will their emotional and social development be affected?

As robots become more integrated into our lives, we should be wary of issues related to privacy and data protection. Many of the tech giants are not exactly known for their careful and respectful attitude toward privacy and data collection. They were responsible for the development and insidious spread of surveillance capitalism[15] as their primary business model: the mining and selling of data on your behavior, data continuously scraped and sucked, often without you knowing. If one of them, or a newcomer in their image, comes to control the domestic robot market, how responsible will they be? Having a (possibly autonomous) mobile device with sophisticated sensors and on-board processing—a robot—in your home, taking you to work, patroling your streets, could increase the amount and quality of data that could be collected by orders of magnitude. Some corporations are not going to pass up that opportunity to snoop on a previously unimagined scale into the deeper details of your life and habits, unless we legislate, control, and regulate.

The most obvious unethical use of robots is in crime. Robots might be reprogrammed from their initial purpose to aid criminals in committing robberies and other felonies. Perhaps more alarming is the possibility of criminals hacking into robots, including driverless cars, to take control in order to perform some illegal act. Security robots could be disabled, or turned against those they are supposed to be protecting; a mobile robot could be used for nefarious surveillance, a vehicle could be suddenly stopped in transit, or worse. Making these options difficult, or easy to detect and override, is an ongoing challenge for the robotics and automotive industries.

Something else that could be regarded as a moral issue is the replacement of human labor with robots. As outlined in Chapter 7, so far it could be argued that robots taking jobs has followed the pattern of previous waves of automation, with new jobs created to fill the holes. However, as more and more low-skilled jobs are handed to robots and then white-collar jobs start to follow, overall employment levels could start to fall significantly. In my opinion, how we, as a society, deal with this, if it comes to pass, will have moral dimensions. Will we knowingly allow inequality to spiral out of control, or will we take another path that values individual dignity?

### Who should be held to account when something goes wrong?

The blame for an accident or illegal behavior caused by the action of a remote-controlled robot would, at least in the first instance, be laid squarely at the feet of the person with the controls. The same would be true of a partly autonomous robot where the human operator is legally responsible at all times, such as for level 3 autonomous vehicles. Dumb, fixed-routine robots are covered by the same general legislation as other industrial machinery. They must be designed, installed, used, and maintained properly. For most robots, accidents due to manufacturing faults or poor design would be treated in the same kind of way as for other machinery and products. But what about accidents or problems caused by decisions made by an autonomous robot, including level 4 and 5 autonomous vehicles? Who or what is responsible? The robot? The owner? The manufacturer? The AI programmer? Some combination of the above? The situation becomes even more complex if the robot's AI learns and develops over time so that it changes from what was initially installed by the manufacturer. Can the accident be put down to poor design of the robot's AI?

Our current legal frameworks, which we use to attribute liability and enforce our systems of ethics so that society can function, urgently need updating to take into account autonomous robots. This is already happening in relation to driverless cars: state-level legislation in some areas of the USA has changed. But there are still many issues that will need clarifying in the future. Autonomous car manufacturers are starting to accept that they will probably be liable for collisions in which their vehicle is at-fault. This of course opens up the whole legal minefield of proving that it was indeed at-fault. Will there be a requirement for the manufacturer to hand over all the data collected by the vehicle? Will there be a comparison with what a human driver should have done in the same circumstances? We will have to accept that machine-learning-driven AI will occasionally make mistakes. AI can never be completely foolproof, just as humans are not perfect. The claim, at least for driverless cars, is that the number of mistakes made by a car's AI will be much less than the number currently made by humans, and overall driving will become much safer.

In the case of hackers taking over robots, or someone programming them to commit crimes, the locus of responsibility and liability shifts, but the law needs to develop in this area.[16]

### Are the ethical issues related to robots different from those that arose with earlier technologies?

In the case of unintelligent or remote-control robots, there are essentially no new issues. As pointed out in the five principles for robotics listed earlier in this chapter, such robots are manufactured machines, and like all other products should comply with relevant laws and regulations to ensure safety and so on. But in the case of autonomous robots, new issues do arise. They are unlike previous technologies; they have the capacity to make their own decisions, to perform their own actions, even if their intelligence is limited. As we have seen, this opens up a whole new seething snake pit of ethical and legal problems.

Bound up with the problems of ethics is the issue of how socially acceptable robots are. The degree of acceptability seems to vary from country to country, with Japan more accepting than most (although, interestingly, there is evidence that this effect is strongest in the older members of the population and drops off in the younger age group[17]).

A startling commentary on robot acceptance is the tale of hitchBOT, a simple humanoid robot created by Canadian academics David Harris Smith and Frauke Zeller.[18] The robot was part art project, part social experiment, designed specifically to ask the question: can robots trust human beings? The experiment was centered around the robot hitchhiking around a country, recording images of its journey while being tracked by embedded GPS. Using mobile phone technology it intermittently posted pictures to its social media accounts. The robot had a cheap, friendly appearance, complete with a smiley LED face. The body was cobbled together from basic everyday items such as a plastic bucket. It was intended to be a robotic traveling companion to any driver who picked it up from the roadside. It could carry out a basic conversation and state facts. Smith and Zeller would leave it by the side of the road at its starting point, with basic instructions written on the back. For a month in summer 2014, hitchBOT successfully traveled right across Canada, going on little adventures and being taken into people's homes. In 2015 a second version was constructed which hitchhiked around Germany for a couple of weeks in February, and then spent three weeks traveling around the Netherlands in June. Starting in mid-July 2015 hitchBOT attempted to hitch across the USA from Boston to San Francisco. Its American journey started well, involving a trip to see a Boston Red Sox baseball game, a boat ride off the coast of Massachusetts, and a visit to Times Square in New York. But it met its end in downtown Philadelphia where it was beheaded and dismembered.

Smith and Zeller stressed that the fact that hitchBOT's violent demise happened in the USA was probably just a coincidence: it was always going to come face to face with vandals sooner or later. But there have been other stories of anti-robot violence in the USA. In Chandler, Arizona, where Waymo has been testing autonomous taxi vans since 2016, the vehicles have had rocks thrown at them, been yelled at, had tires slashed, been chased, and been forced off the road. One had a gun waved at it. There have been incidents where delivery robots were knocked over, or security robots sabotaged—remember the SFSPCA robot guard from Chapter 1? The USA is at the forefront of introducing autonomous robots into the workplace, there are more experimental robots out in the open than anywhere else, and forecasts for the number of jobs that will be taken by robots are particular high there. So perhaps the violence is not a coincidence. Or perhaps it is a general primal fear of "the other" that wells up in some people.

# 9

# ROBOT FUTURES

*How might we be using robots in the relatively near future? say in fifty or sixty years' time?*

Of course, forecasting technological futures is a mug's game. Technology often advances at a steady pace, so it is possible to cautiously extrapolate. But there are also sudden bursts of rapid development. These bursts rarely last as long as people imagine they will, and many problems turn out to be much harder than first thought. Conversely, there are unexpected game-changing discoveries that fly in from left-field and throw everything into the air. Hence, all speculations in this chapter should be taken with a large pinch of salt. As the saying goes, the future is a door that cannot be opened until it opens.

So here I attempt to sketch plausible robot futures, some positive, some negative, some in the near future, and some in the far future. I do so using short descriptive scenarios, rather than fact-based argument. These vignettes echo some of the issues developed in earlier chapters. I hope this approach will help to further elaborate a number of the main scientific and social challenges that must be faced.

To answer the question about the relatively near future, let's imagine the following scenarios. The first highlights some of the positive contributions robots *might* make, and the second illustrates some potential problems.

\*\*\*

Mary slipped on her super-lightweight exoskeleton suit. It quickly adapted its shape to fit her snugly. She could hardly feel it was there, but it was a tremendous boost to her mobility on the days she had a lot to do. It gave her the kind of feeling she'd had from the motorized bicycle that had effortlessly whizzed her up hills years ago. Today she was off to visit her daughter, Christina, to discuss arrangements for her 105th birthday party. She didn't want any fuss, but Christina had insisted. She'd need to book a roboCab. There were numerous ways of doing this, what with nearly all devices having voice interfaces with access to every app under the sun, but she liked to book by talking to one of her robot vacuum cleaners. She had two, one named Grey and the other Walter. Before she'd retired, she'd worked as a roboticist for the company that produced them. She was pretty sure some of her software gems were still in there somewhere. Luckily, her daughter lived within the same traffic control zone as her so she could take an autonomous electric cab all the way. Driverless cars were still not allowed outside the control zones, so for longer journeys she'd swap-over halfway to one of the new luxury coaches everyone was encouraged to take. Occasionally, she'd drive herself. She was one of the last to still own a private semi-autonomous vehicle. Most people disapproved, but old habits and all that. She'd had to donate her lovingly preserved classic Jaguar E-type to a museum a few years ago when manual petrol vehicles were outlawed on public roads. She had to admit that traffic was much calmer and quieter than it used to be, and of course pollution had been slashed. She knew it was wrong, but she found herself longing for the days when she'd break the speed limit on the back roads on the way home.

She booked the roboCab to come in an hour which gave her time to call on a serviceBot. There were a few shared between the houses in her little neighborhood. The buildings had been specially built to allow the machines easy access. She admired the robots' sleek design. Various retractable wheel mechanisms, combined with limbs, and moveable arrays of sensors made them amazingly flexible. But she understood their limitations better than most. They were not fully autonomous; a remote human operator occasionally came into play for some tasks. She had had to sign a form allowing the robot company access to live sensor feeds in certain circumstances while their robots were in her house. But the serviceBots could turn their three hands to most simple chores she asked for help with. They did some of the messier cleaning jobs for her and some simple cooking, and helped her into the shower on days her rheumatism was playing up. The injections of nanobots into her bloodstream worked wonders, but it wasn't a cure. The tiny, simple organic robots had one job: to precisely target her medication before breaking down into harmless waste that was filtered out. Most of all, she liked to chat to the serviceBots. Their conversational ability amused her. It was impressive, but she liked to catch them out in nonsensical responses, proving they had lost the thread. Yep, you still can't pass the Turing test, she murmured.

***

Billy was waiting in the queue near the truck interchange. It was 5:00 am and an icy wind whipped his face. He had hardly eaten for two days and was starting to feel desperate. He had a nasty cough he couldn't seem to shake off. This was where all the drivers came to try and get a job taking one of the trucks into the city or to the warehouses off the control route. Knots of truckers were strung out across the verge, chatting and fretting. They generally only got a couple of hours' work, but it was something. The giant trucks went in driverless convoys up to this point, but needed humans to guide them on the final more difficult stretches. There were rumors that legislation might be passed soon to allow them to go driverless all the way. He hoped not. There were so few driving gigs left. He always got here as early as possible, but often there weren't enough slots to go round the increasingly unruly crowd of drivers. They had contracts with a company that serviced the huge warehouses and supermarkets, but they weren't worth a damn. The days when he'd had as much work as he could handle were long gone. Often there were delays when one of the trucks in the convoy broke down or reported an error. The whole convoy had to wait for a maintenance crew to arrive, or spend time reconfiguring when the dud truck was abandoned.

The move toward fully autonomous trucks in all traffic zones had looked certain until that driverless 18-wheeler they'd been testing somewhere in California took a wrong turn and smashed into an office block, destroying half the street as it went into recovery mode. Apparently it just kept going back and forth as if it was trying to execute a three-point turn or something. They say it took out the HQ of the AI company that just happened to be the main rival to the one that supplied the truck's software. At least 30 were killed. That's what they were saying, anyway. It seemed like a conspiracy theory at first, but all the main news channels were carrying the story, and they were reporting that some AI engineer who had worked at both companies had been arrested.

It hadn't been so bad when the driving jobs first started to disappear. Back then there were still plenty of openings in the big warehouses out to the west of the town. But now all those jobs had been taken over by robots. Loading, unloading, and moving stock around the floor were already mainly done by robots when Billy had worked there. Now all the picking and packing jobs were fully automated too. An army of thousands of robots ceaselessly reached, plucked, stacked, wrapped, pushed, and pulled. Their hisses, clunks, and beeps echoed through the vast, arid metal barns. The security guards were all robots now, and they were multiplying. There were just a few human supervisors left and a team of on-site engineers to look after the robots.

Billy drove trucks to and from the warehouses maybe twice a week. On every trip he never saw a human or heard a human voice, just the canned commands of the robots telling him to stay in his cab while they loaded or unloaded. The home-delivery routes were largely autonomous, so they only wanted fit young van drivers. Their main job was to run from van to door with the goods to maximize the delivery rate. That ruled him out. Flying delivery drones were being used more now as people started to install secure land-and-drop units at the back of their properties.

When the first big wave of job losses came, there was still support and he'd been offered retraining. He'd tried to learn to code, but there had only been two sessions and the rest was online. It wasn't for him. He just couldn't get the hang of it at his age, what with his lack of schooling. Now there was no help at all. The mega tech corporations that owned every-thing, including the politicians, had seen to that. They called it optimizing their legal tax obligations. He called it something else. Someone had told him they had teams of analysts trying to work out how many jobs they could automate and how many they should help create so as not to destroy their own markets. If this was true, he knew that as long as the profits kept increasing, people like him would never figure on the positive side of that equation.

One of the other drivers was trying to get Billy interested in the Anti-Robot League. Now that white-collar jobs were starting to be offloaded to robots, the League was growing again. He reckoned that already maybe 20% of service and support jobs were done by robots: cleaning, some care and basic nursing, reception staff, some retail staff, waiters, even bar staff. If that trend continued, and spilled over to higher-skill areas, there was going to be big trouble. The League was classified as a terrorist organization by the government; it mainly organized protests, but there had been some direct action against a robot construction plant and a robot ship-ment coming in from overseas. Only machines had been hurt. In response, the tech corporations were stepping up security. Huge threatening robots were patrolling day and night. They were pushing to legalize arming the robots. Non-lethal force was already allowed if the machines were attacked; now they wanted their robots locked and loaded.

Maybe he would go to the next League meeting.

*\*\*\**

### How might we be using robots in the far future? Say in five or six hundred years' time?

Guessing what might happen that far in the future is of course impossible. But let's imagine the following highly speculative scenarios. The first illustrates some potential problems, particularly in relation to dehumanizing effects, while the second highlights some of the positive contributions robots might make. They give an indication of the type of timescale I would bet on for major progress toward powerful general intelligent robotics—not decades but centuries (don't forget we have been trying for the best part of a century already, and are only just starting to achieve powerful but very narrow, very specific AI).

*\*\*\**

"Brilliantly put, as ever. You have such clarity of thought, such a beautiful way of cutting to the heart of the matter," said Henry.

Cerise felt a surge of irritation. She walked to where Henry stood smiling vacantly. His perfect hair, his too taut skin, his evenly modulated tones, they all annoyed her. She waved her hand near his right ear.

"Administration access code 121," she said sharply.

Henry's ear glowed green, indicating voice and bio-scan security were successfully passed.

"Personality type edit. Turn obsequiousness to level 2. Turn strong and silent to 7," ordered Cerise.

Henry's ear flashed amber twice. He crossed the room and sat in a chair, staring moodily out the window.

She turned and spoke to him, "I'm thinking of visiting Anton tonight. Do you want to come?"

Henry ignored her. She sighed and smiled to herself. Cerise had used the roboPartner for about five or six years now and was starting to tire of its limitations. Everything was, well, just that bit too obviously artificial. From its looks to its conversation. In some ways she preferred Henry to Anton, her human bio-partner and father of her children, but she was starting to wonder about spending more time with Anton.

The unmistakable hum of an electric aircraft landing came from out front and the main door gently opened as the children rushed in. They were greeted by the robot maid, who laughed as they hugged her and excitedly told her about their day. They ran upstairs to play with her, exchanging blank glances with their mother as they passed her in the corridor. She liked to say that the maid had been expensive but well worth it. The kids had spent most of their time with her since birth. Sometimes one of the serviceBots helped, but it was mainly the indefatigable maid who did everything for them. Cerise was nervous about what might happen if the robot ever broke down, so she had it serviced every three months. Its micro quantum compute engine was more advanced than Henry's, and its neural substrates were supposed to last decades without degradation as long as they received the right nutrients. The robot knew more about that than her and took care of its own food and fuel.

Cerise started to program a route for the autonomous aircraft; she'd decided to visit a friend in the next compound. The machine checked conditions and morphed into a bird-like flapping-wing configuration, which was more efficient and maneuverable for the planned trip. As she walked outside she could hear dull explosions coming from somewhere beyond the perimeter.

There were supposed to be other humans living in the BurntLand, members of the underclass who had somehow survived the Catastrophe. She'd never seen any but was happy that the perimeter robot guards kept them away. When they flew over the desert areas their craft became invisible and scanned the ground relentlessly, just in case. Outsiders were not welcome. They had everything they wanted, thanks to robot labor, but that didn't mean there was enough to share.

When she touched down, Cerise was met by two humanoid servant robots who bowed and sprayed her with a refreshing mist to help with the searing heat. She started screaming about her shoes and began kicking and punching one of the robots. She swung her bag at it, aiming the sharp corner at its eyes. She could see Serge sauntering down from the house carrying a pistol. Cerise explained that the idiot robot had got water on her new shoes. He joined in the beating, smashing its face with the gun grip. He then fired into its head and twisted off an arm before shooting at its legs. Cerise and Serge laughed and walked toward the house.

The other robot picked up its companion and inspected the damage. Neurons fired and quantum interfaces crackled. Almost instantly it had diagnosed all damage and planned the necessary repairs. Its eyes narrowed as it watched the two humans disappear into another chemically assisted hour, as empty as the last.

\*\*\*

This morning the tutor was leading an immersive history experience for an under-ten group in the new education building. They were learning about the type of work done by their ancestors hundreds of years ago. As they wandered through the super-definition holographic displays of twentieth- and twenty-first-century offices, factories, and other workplaces, many of the students found it hard to grasp how people had spent so much of their lives in such monotonous, or stressful, or unrewarding occupations that harmed their health. The display had been set up and was being coordinated by a robot assistant.

One of the children pointed at the friendly looking semi-humanoid robot busily moving icons around a complicated-looking control panel. "Didn't they have robots by then?"

The tutor explained that it took a long time for the robots to reach the levels they all took for granted, and almost as long for their society to organize itself to make best use of them.

Like many adults in this part of the world the tutor had several registered occupations that were scheduled in a flexible way (he was also a musician and a medical practitioner). He'd decide each month how many hours he'd like to spend on each and was free to use the remaining time on leisure, education, or other projects. Other people preferred to concentrate on a single role. The large robot workforce and advanced production processes meant that everyone had a guaranteed comfortable income, as long as they made some kind of contribution if they were able. It was possible to earn more—up to a maximum—from entrepreneurial efforts or by taking on extra responsibility. But most people did what they did because they found it interesting or rewarding, because they enjoyed it.

The tutor introduced a historian from the university. She described to the class how it had taken many years, and some violent struggle, to get to where they were now, and how other parts of the world still used the old models. She was referring to the way their society integrated with robots. Robots were seen as tools to benefit all. Robots were used to help and improve human life, but never to replace human activity, unless better alternatives were available to the humans. She explained how this attitude, which was the basis of their stability and prosperity, had not always held sway. No one had yet succeeded in developing general human-level AI, although there were many impressive systems out there that outperformed humans in certain areas. She thought that was probably a good thing—it helped to keep a perspective on robots as human-created tools. She reminded them that for roles that could be done by either a robot or a human, humans had a legal right to the role, if they wanted it. Interchanging roles between humans and robots were common—no doubt they'd all been taught by robot tutors on some days—and learning to think of robots as helpful, unthreatening co-workers was a useful achievement of their society.

Afterward they went outside for a break. They watched the pulsing, mesmerizing colors of a swarm of robot insects heading off toward the agri-zone. The tiny robots had many uses: redirecting crop pests to areas where they would do no harm, managing pollination, and collecting reams of data on every aspect of crop health, as well as local environmental conditions. Above them, elegant robot birds with startling plumage were gliding over the fields on the look-out for larger pests, as well as directing ground-based irrigation robots to the best target areas.

Some of the teenagers nearby were chatting about the Borg Games. The cyborg athletics championships were to be held at the sports complex in a few days. There was talk of skullduggery from one of the main teams: outlawed genetic engineering practices and banned neural implants that deadened pain. The team owner denied all allegations and challenged the authorities to test his cyborgs. They'd find only legal enhancements, he claimed.

"Isn't that the same guy who reckons he is going to upload his consciousness onto a compute-engine next year?" asked one of the crowd.

"Yeah, that's him. Apparently he's been saying that for the past 30 years."

***

Our robot futures are as yet unformed. It is up to us collectively to make the best of them.

# NOTES

**Chapter 1**

1. Green, A. (2017) Security robot that deterred homeless encampments in the Mission gets rebuke from the city, *San Francisco Business Times Dec 8*. Bromwich, J. (2019) Why do we hurt robots? *New York Times Jan 19*.

2. Green, A. (2017) The robots are coming to San Francisco but now they have rules, *San Francisco Business Times Dec 6*.

3. Mormann, F. et al. (2011) A category-specific response to animals in the right human amygdala, *Nature Neuroscience* **14**:1247–1249.

4. Oztop, E., Franklin, D., Chaminade, T., and Cheng, G. (2005) Human–humanoid interaction: is a humanoid robot perceived as a human? *International Journal of Humanoid Robotics* **2**(4):537–559.

5. Broadbent, E. (2017) Interactions with robots: the truths we reveal about ourselves, *Annual Review of Psychology* **68**:627–652.

6. Broadbent, E., MacDonald, B., Jago, L., Juergens, M., and Mazharullah, O. (2007) Human reactions to good and bad robots, In *Proceedings of the 2007 IEEE/RSJ International Conference on the Intelligent Robots and Systems (IROS)*, pp.3703–3708. Spatola, N. et al. (2018) Not as bad as it seems: when the presence of a threatening humanoid robot improves human performance, *Science Robotics* **3**(21):eaat5843.

7. Mori, M. (2012) Translated by MacDorman, K.F. and Kageki, N. The uncanny valley, *IEEE Robotics and Automation* **19**(2):98–100 (originally appeared in Japanese in 1970). Mathur, M. and Reichling, D. (2016) Navigating a social world with robot partners: a quantitative cartography of the uncanny valley,

*Cognition* **146**:22–32. There is still debate about whether or not the effect really exists, and, if it does, where the valley lies on the scale of degree of human likeness, and if it is age and culture dependent.

8. International Federation of Robotics, World Robotics Report 2016, https://ifr.org/ifr-press-releases/news/world-robotics-report-2016 [retrieved Jan 2019].

9. iRobot Reports Record Fourth-Quarter and Full-Year Revenue, iRobot Press Release, http://media.irobot.com/2018-02-07-iRobot-Reports-Record-Fourth-Quarter-and-Full-Year-Revenue [retrieved Jan 2019].

10. IFR Executive Summary World Robotics Service Robots 2017, https://www.ifr.org/.../press/Executive_Summary_WR_Service_Robots_2017_1.pdf [retrieved Jan 2019].

**Chapter 2**

1. See e.g., Hanson, D. (2011) Why we should build humanlike robots, *IEEE Spectrum, April 1*.

2. Du, R. et al. (2015) Robot Fish: Bio-inspired Fishlike Underwater Robots, In *Springer Tracts in Mechanical Engineering*, Cham: Springer.

3. Robot ships have arrived . . ., http://www.youngship.com/youngship-international/news/robot-ships-have-arrived-and-nobody-told-us/ [retrieved May 2019]. How Europe's busiest port is helping make autonomous ships a reality, https://www.nbcnews.com/mach/science/how-europe-s-busiest-port-helping-make-autonomous-ships-reality-ncna978041 [retrieved May 2019].

4. Rus, D. and Tolley, M. (2015) Design, fabrication and control of soft robots, *Nature* **521**(7553):467–475.

5. There are now various companies, such as Suitable Technologies Inc., marketing commercial telepresence robots.

6. Heavy-duty industrial robots tend to use very accurate sensors and actuators to do their job efficiently and comply with tight safety rules.

7. In Figure 2.2, for simplicity the information flow from sensors to control system to actuators is shown as one way all the way. This is often the case, but in more sophisticated robots it can be more complex. For instance, the control system might also directly feed back to the sensors, altering their properties, or self-regulating

mechanisms in the actuators might directly feed back to the control system, changing its responses.

8. If both wheels are driven forward/backward at the same speed, the robot will move ahead/backward in a straight line. If the right motor is driven faster than the left, the robot will move in a curve toward the left (and vice versa). Depending on the exact combination of motor speeds the robot can move at various speeds in gentle or tight curves (forward and backward). It will spin on the spot when one wheel goes forward and the other goes backward at the same speed.

9. We would now need extra sensors to tell if the robot is stuck, i.e., not moving.

10. The basic "time of flight" calculation for determining the distance from a rangefinder is 2.D = c.t, where D is the distance (hence 2.D is the total distance traveled by the laser pulse), c is the speed of light, and t is the time taken for the return trip from senor to object and back again.

11. The cerebral cortex is by far the largest part of the human central nervous system, taking up nearly 80% by volume; about a third of it is concerned with vision. (Source: Swanson, L. (1995) Mapping the human brain: past, present and future, *Trends in Neuroscience* **18**(11):471–474.)

12. Beer, R., Quinn, R., Chiel, H., and Ritzmann, R. (1997) Biologically-inspired approaches to robotics, *Communications of the ACM* **40**(3):30–38.

13. A human stride involves swinging a leg out to catch ourselves as we fall forward; we unconsciously do this from step to step and miraculously move forward in a rather elegant way. (Wang, Y. and Srinivasan, M. (2014) Stepping in the direction of the fall: the next foot placement can be predicted from current upper body state in steady-state walking, *Biology Letters* **10**(9):20140405). See e.g., https://www.youtube.com/watch?v=JlRPICfnmhw [retrieved Feb 2019] for a video of Honda's famous Asimo robot walking in its distinctive crouched position.

14. Hauser, H., Ijspeert, A.J., Fuchslin, R.M., Pfeifer, R., and Maass, W. (2011) Towards a theoretical foundation for morphological computation with compliant bodies, *Biological Cybernetics* **105**(5–6):355–370. Johnson, C., Philippides, A., and Husbands, P. (2019) Simulating soft-bodied swimmers with particle-based physics, *Soft Robotics* **6**(2):263–275.

15. Pfeifer, R. and Bongard, J. (2006) *How the Body Shapes the Way We Think,* Cambridge, MA: MIT Press. Di Paolo, E., Buhrmann, T., and Barandiaran, X. (2017) *Sensorimotor Life: An Enactive Proposal,* Oxford: OUP.

16. See https://www.rethinkrobotics.com/.

17. Smithsonian, Pepper the robot, https://www.si.edu/visit/pepper [retrieved April 2019]. The Pepper robots are developed by the Franco-Japanese company Softbank Robotics.

18. See a Pepper robot in action in a restaurant at https://www.youtube.com/watch?v=VkBTMugJ7wU [retrieved April 2019].

19. Barbash, G. and Glied, S. (2010) New technology and health care costs—the case of robot-assisted surgery, *The New England Journal of Medicine* **363**(8):701–704. Hyldgård, V. et al. (2017) Robot-assisted surgery in a broader healthcare perspective: a difference-in-difference-based cost analysis of a national prostatectomy cohort, *BMJ Open* **7**:e015580.

## Chapter 3

1. Horáková, J. and Kelemen, J. (2008) The robot story: why robots were born and how they grew up, In Husbands, P., Holland, O., and Wheeler, M. (Eds) *The Mechanical Mind in History,* Cambridge, MA: MIT Press, pp.283–306 .

2. Truit, E. (2017) In whose image? Ancient and medieval automata, In Russell, B. (Ed) *Robots: The 500 Year Quest to Make Machines Human,* London: Scala Arts and Heritage Publishers, pp.33–47.

3. Caused by overblowing some high notes due to the limitations of the android's lips and mouth.

4. Wood, G. (2002) *Living Dolls: A Magical History of the Quest for Mechanical Life,* London: Faber & Faber.

5. For a detailed discussion of how such automata reflected and influenced enlightenment views see Schaffer, S. (2001) Enlightened automata, In Clark, W., Golinski, J., and Schaffer, S. (Eds) *The Sciences in Enlightened Europe,* Chicago: University of Chicago Press, pp.126–165.

6. The Silver Swan is housed in the Bowes Museum, Barnard Castle, Teesdale, County Durham, England. To preserve its mechanisms it performs only once a day at most.

7. For instance in 1742 Vaucanson wrote a pamphlet on the inner workings of his machines, which was translated into English soon after its appearance in French: Vaucanson, J. (1742) *An*

*Account of the Mechanism of an Automaton*, London: Printed by T. Parker, and sold by Mr. Stephen Varillon. The transcript of a lecture he presented to the Royal Academy of Science in Paris, it describes the mechanisms of the flute player, a tabor player, and the defecating duck.

8. See e.g., Levitt, G. (2007) *The Turk, Chess Automaton*, Jefferson, NC: McFarland & Co.

9. Mitchell, S. (1857) The last of a veteran chess player, *The Chess Monthly* (reprinted in G. Levitt, 2007).

10. For the most complete account of Eric (and the later George) see this article by two of WH Richards' great-granddaughters: Richards Lever, A. and Richards, D. (2017) Robots in the family: Captain WH Richards and his mechanical men, In Russell, B. (Ed) *Robots: The 500 Year Quest to Make Machines Human*, London: Scala Arts and Heritage Publishers, pp.70–83; see also Reuben Hoggett's excellent cyberneticzoo.com page on Eric which includes transcribed newspaper and magazine articles from the 1920s, http://cyberneticzoo.com/robots/1928-eric-robot-capt-richards-english/ [retrieved August 2019].

11. A short clip of Eric performing can be seen at https://www.youtube.com/watch?v=lLmohGA19Ek [retrieved August 2019].

12. *Chronicle Telegram*, Nov 26, 1928, p.14; this statement does not preclude the use of some other recording technology of the time, such as phonograph cylinders.

13. For further reading on Elektro see e.g., Marsh, A. (2018) Elektro the Moto-Man had the biggest brain at the 1939 World's Fair, *IEEE Spectrum Sept 28*.

14. For a clip of a cheesy Westinghouse promo of Elektro performing at the World Fair see https://www.youtube.com/watch?v=AuyTRbj8QSA [retrieved August 2019].

15. Some of the biographical details of Grey Walter are taken from an unpublished autobiography written by Grey, held in the private papers of the Walter family who generously gave the author access. Many thanks to Natasha Walter for that.

16. Whether for his frequent womanizing, his indirect involvement with the Cambridge spies, or his sometimes over-enthusiastic interactions with the media.

17. Walter, W.G. (1943) An automatic low frequency analyser, *Electronic Engineering* **16**:8–13. Walter, W.G. (1943) An improved low frequency analyser, *Electronic Engineering* **16**:236–238.

18. Walter, W.G. (1975) *Grey Matter: The Life Story of a Brain Scientist*, unpublished memoir, private papers of Walter family, Chapter 14, pp.16–17.

19. He used analog rather than digital computers—devices using continuous analog electronics to make specific calculations, very different to today's general-purpose, programmable digital electronic computers.

20. Craik, K.J.W. (1943) *The Nature of Explanation*, Cambridge: Cambridge University Press.

21. Walter, W.G. (1953) *The Living Brain*, London: Norton, p.125.

22. The exact dates of completion are unknown. de Latil, P. (1956) *Thinking by Machine*, London: Sidgwick and Jackson, p.209 states that Elmer was finished in 1948 (as do other contemporaneous sources) and Elsie a few months later, based on discussions with Grey in the early 1950s. A newspaper article from September 1949 reports on a demonstration of both, so they were completed before then (*Daily Herald*, Sept 21, 1949, p.3, "Elsie the tortoise does jig").

23. The Festival of Britain was a large national exhibition spread over several sites in London. It was designed to celebrate national achievements in science, technology, and the arts, and aimed to give the nation a feeling of recovery from the devastation of the recent war.

24. See an old (*circa* 1950) newsreel of the tortoises working here: https://www.youtube.com/watch?v=lLULRlmXkKo [retrieved Sept 2019].

25. Walter, W.G. (1953) *The Living Brain*, London: Norton, p.128.

26. Brooks, R.A. (2002) *Flesh and Machines: How Robots Will Change Us*, New York: Pantheon Books, p.27.

27. Electronic valves (or tubes in the USA) are devices in which current flows in a vacuum between electrodes housed in a glass tube. Up until the mid-1960s they were used in a very wide range of applications of electronics, including amplification. After that they were largely replaced by transistors and solid-state electronics.

28. Walter, W.G. (1950) An imitation of life, *Scientific American* **182**(5):42–45.

29. de Latil, P. (1956) *Thinking by Machine*, London: Sidgwick and Jackson, pp.240–241. Another good source of information on Philidog is http://cyberneticzoo.com/precyber/

1928-phil-the-radio-dog-a-k-a-philidog-piraux/ [retrieved
Sept 2019]. A similar phototropic device, which may have
influenced Piraux, had been built in the USA with earlier
photoelectric technology as long ago as 1912 by John Hays
Hammond and Benjamin Miessner, who in turn were probably
inspired by physiologist Jacques Loeb's work on animal
tropisms from the turn of the century (*Scientific American*, June
14, 1919, p.376; see also Cordeschi, R. (2002) *The Discovery of
the Artificial*, Berlin: Springer, for details of various other early
robots).

30. Holland, O.E. (1996) Grey Walter: the pioneer of real artificial life,
    In Langton, C. and Shimohara, K. (Eds) *Proceedings Artificial Life
    V*, Cambridge, MA: MIT Press, pp.33–41.

31. Wiener, N. (1948) *Cybernetics, or Control and Communication in the
    Animal and the Machine*, Cambridge, MA: MIT Press.

32. Husbands, P. and Holland, O. (2008) The Ratio Club: a hub of
    British cybernetics, In Husbands, P., Holland, O., and Wheeler,
    M. (Eds) *The Mechanical Mind in History*, Cambridge, MA: MIT
    Press, pp.91–148.

33. Wiener, N. (1954) *The Human Use of Human Beings: Cybernetics and
    Society*, 2nd edn, New York: Doubleday, pp.163–167.

34. Nilsson, N.J. (Ed) (1984) *Shakey the Robot*, Technical Note 323,
    Menlo Park, CA: AI Center, SRI International.

35. Moravec, H. (1987) Sensing versus inferring in robot control,
    Informal Report, www.frc.ri.cmu.edu/~hpm/project.archive/
    robot.papers/1987/sense.ltx [retrieved Sept 2019], p.1.

36. Brooks, R.A. (1999) *Cambrian Intelligence: The Early History of the
    New AI*, Cambridge, MA: MIT Press.

37. Raibert, M.H. (1986) Legged robots, *Communications of the ACM*
    **29**(6):499–514.

## Chapter 4

1. Brooks, R.A. (1986) A robust layered control system for a mobile
   robot, *IEEE Journal of Robotics and Automation* **2**(1):14–23.

2. Pomerleau, D. (1989) ALVINN: An Autonomous Land Vehicle
   in a Neural Network, In Touretzky, D.S. (Ed) *Proceedings of
   Conference on Advances in Neural Information Processing Systems* 2,
   Burlington, MA: Morgan Kaufmann, pp.305–313.

3. Thorpe, C., Herbert, M., Kanade, T., and Shafer, S. (1988)
   Vision and navigation for the Carnegie Mellon NAVLAB,

*IEEE Transactions on Pattern Analysis and Machine Intelligence* **10**(3):362–373.

4. This transformation is often a matter of squashing the summed input values into a narrow range (usually 0 to 1). The logistic function, $f(x)=1/(1 + e^{-kx})$, was used by ALVINN, where x is the summed inputs. Another simple transformation sometimes used is: if the sum of all inputs is greater than some pre-defined threshold, then the output is 1; otherwise it is 0.

5. Named after Donald Hebb, a Canadian psychologist, who first proposed this learning rule in relation to real neurons in his influential 1949 book, *The Organization of Behavior*, Wiley.

6. These kinds of methods have become particularly popular since a team at Deep Mind showed how they could be used to enable an ANN to learn to play ATARI games from scratch (Mnih, V. et al. (2015) Human-level control through deep reinforcement learning, *Nature* **518**(7540): 529–533).

7. There was actually one further output that measured differences in light intensities; this was fed back to an extra input and helped to deal with changing lighting conditions and shadows.

8. Back-propagation learning was used: Rumelhart, D., Hinton, G., and Williams, R. (1986) Learning representations by back-propagating errors, *Nature* **323**(6088):533–536.

9. Smith, R., Self, M., and Cheeseman, P. (1990) Estimating uncertain spatial relationships in robotics, In Cox, I. and Wilfon, G. (Eds) *Autonomous Robot Vehicles*, Berlin: Springer, pp.167–193. Durrant-Whyte, H. and Bailey, T. (2006) Simultaneous localisation and mapping (SLAM): part I the essential algorithms, *Robotics and Automation Magazine June*.

10. Thrun, S., Burgard, W., and Fox, D. (2005) *Probabilistic Robotics*, Cambridge, MA: MIT Press.

11. Wiener and several other cyberneticians, including Albert Uttley, Jack Good, and Alan Turing, developed methods related to probabilistic inference for refining signals or "models" based on observed data. Some of these methods became known as "filtering," because their primary goal was to filter out noise in an uncertain signal. These were further developed by control theorists as part of techniques to automatically control machines in, for example, chemical plants.

12. Thrun, S. et al. (2006) Stanley: the robot that won the DARPA Grand Challenge, *Journal of Field Robotics* **23**(9):661–692.

13. LIDAR was developed as a surveying method for building digital 3-D representations of a target (often an area of landscape). It uses time of flight of reflected pulsed (and often scanned) laser light to measure distances from points on the surfaces making up the target.

14. Davis, A. (2019) This Lidar is so cheap it could make self-driving a reality, *Wired July 11*.

15. For instance, Affectiva, an MIT Media Lab spinoff, in 2018 launched an AI system for monitoring drivers' moods and emotions. Jaguar Land Rover is investigating similar systems that change the internal environment (lighting, temperature, even music) in response to mood.

16. Sage, A. (2018) Waymo unveils self-driving taxi service in Arizona for paying customers, *Reuters Dec 5*. In this service, available in a small well-defined area to pre-approved subscribers only, a human driver sits behind the wheel just in case.

17. For example, Miura, H. and Shimoyama, I. (1984) Dynamic walk of a biped, *International Journal of Robotics Research* 3(2):60–74.

18. Brooks, R.A. (1989) A robot that walks; emergent behaviors from a carefully evolved network, *Neural Computation* 1(2):253–262.

19. Beer, R.D., Chiel, H.J., Quinn, R.D., Espenschied, K., and Larsson, P. (1992) A distributed neural network architecture for hexapod robot locomotion, *Neural Computation* 4(3):356–365.

20. Nolfi, S., Bongard, J., Floreano, D., and Husbands, P. (2016) Evolutionary Robotics. In Siciliano, B. and Khatib, O. (Eds) *Springer Handbook of Robotics, 2nd edn*, Chapter 76, Berlin: Springer, pp.2035–2067.

21. Turing, A.M. (1950) Computing machinery and intelligence, *Mind* **59**:433–460.

22. Lipson, H. and Pollack, J. (2000) Automatic design and manufacture of robotic lifeforms, *Nature* **406**:974–978.

23. Baddeley, B., Graham, P., Husbands, P., and Philippides, A. (2012) A model of ant route navigation driven by scene familiarity, *PLoS Computational Biology* **8**(1):e1002336.

24. For instance, a popular safety method is triple modular redundancy where safety-critical modules operate in triplicate. The outputs are checked and if they differ (presumably because one has a fault) the majority output is used (i.e., that of the other two copies). This is very expensive and has a potentially fatal

flaw: what if there is a fault in the checking circuitry?—who guards the guard? (Garvie, M. (2005) *Reliable Electronics Through Artificial Evolution*, Doctoral Thesis, University of Sussex).

25. Cha, E., Kim, Y., Fong, T., and Mataric, M. (2016) A survey of nonverbal signaling methods for non-humanoid robots, *Foundations and Trends in Robotics* **6**(4):211–323. Mavridis, N. (2015) A review of verbal and non-verbal human–robot interactive communication, *Robotics and Autonomous Systems* **63**(1):22–35.

## Chapter 5

1. There seems to be another big-budget adaptation of *War of The Worlds* every few years; for instance, in November 2019 the BBC launched an ambitious new mini-series based on the book.
2. These themes are explicitly discussed in the novel, through the voices of various characters.
3. Haynes, R. (1979) *H.G. Wells: Discover of the Future*, London: Macmillan.
4. The robot costume was designed by Walter Schulze-Mittendorff and worn by the actor Brigitte Helm. For much of the film the robot actually takes human form, via a memorable transformation scene, appearing with the likeness of Maria, the saintly protagonist who is trying to bring the workers and overlords together. The fake "robot Maria" is used by the mad scientist to create chaos and dissent in Metropolis with the intent of bringing it crashing down. Eventually, good prevails in the rather corny plot, with the leader of the workers and the head of the city joining hands in the final frame.
5. Huxley, A. (1932) *Brave New World*, 2nd edn, London: Chatto and Windus, p.18.
6. Grey's close friend Victor Rothschild brought Huxley to see Grey's experimental work. Grey Walter (1975), from Chapter 9 of his unpublished autobiography, *Grey Matter: The Life Story of a Brain Scientist*, private papers of the Walter family.
7. For instance: Justus, K. et al. (2019) A biosensing soft robot: autonomous parsing of chemical signals through integrated organic and inorganic interfaces, *Science Robotics* **4**:31.
8. Kriegman, S., Blackiston, D., Levin, M., and Bongard, J. (2020) A scalable pipeline for designing reconfigurable organisms, *Proceedings of the National Academy of Sciences* **117**(4):1853–1859.

9. Clarke's first novel, *Against the Fall of Night*, published in 1948, was set in the year 10 billion CE.

10. Boston Dynamics, founded by MIT roboticist Marc Raibert in 1992, build some of the most advanced mobile legged robots available https://www.bostondynamics.com/. In various interviews Charlie Brooker has cited videos of Boston Dynamics robots as direct influence on *Metalhead* (e.g., Hibberd, J. (2017) Black Mirror creator explains that "Metalhead" robot nightmare, *Entertainment Weekly Dec 29*).

11. Released in September 2020, developed by CD Projekt RED.

12. For example Yoshua Bengio's AI Horizons keynote at the AI Research Week Colloquium, September 2019, to name but one of many.

13. For example in his 1942 story *Runaround*, published in the March 1942 issue of *Astounding Science Fiction*.

14. Walter, W.G. (1956) *Further Outlook*, London: Duckworth (renamed *The Curve of the Snowflake* for the USA edition). Although the novel contains some interesting ideas, fiction was not Walter's forte.

15. The original Turing test, proposed in 1950 by Alan Turing, was a simple (and rather imprecisely defined) operational test to decide whether or not a machine (possibly a program running on a computer) was intelligent. The idea was for an "interrogator" to hold a conversation via a teletype terminal. At the other end, unseen, there were two further "players," a machine and another human. If the interrogator could not tell which was the machine, after asking both multiple questions over some reasonable length of time, then the machine could be considered intelligent. (Turing, A. (1950) Computing machinery and intelligence, *Mind* LIX (Issue 236):433–460.

16. *2001: A Space Odyssey*, MGM, 1968, screen play by Stanley Kubrick and Arthur C. Clarke, directed by Stanley Kubrick.

17. I refer to HAL as "he," as do the other characters in the film, because the machine was given a male personality. Strictly HAL is an it.

18. Interview with Stanley Kubrick in Gelmis, J. (1970) *The Film Director as Superstar,* Garden City, NY: Doubleday.

19. Skarda, C. and Freeman, W. (1987) How brains make chaos in order to make sense of the world, *Behavioral and Brain Sciences* **10**:161–195. Shim, Y. and Husbands, P. (2019) Embodied

neuromechanical chaos through homeostatic regulation, *Chaos*
**29**(3):033123.

## Chapter 6

1. Ulman, S. (1958) John Von Neumann, 1903–1957, *Bulletin of the American Mathematical Society* **64**(3), part 2:1–49.
2. Good, I.J. (Ed) (1962) *The Scientist Speculates: An Anthology of Partly-Baked Ideas*, New York: Basic Books. Contributors included cybernetics and AI pioneers Oliver Selfridge, Gordan Pask, and Marvin Minksy, biologists John Maynard Smith, Lionel Penrose, and C.H. Waddington, writers Arthur C. Clarke and Isaac Asimov, mathematicians J.E. Littlewood and George Polya, philosopher Michael Polanyi, and many more.
3. Good, I.J. (1962) The social implications of artificial intelligence, in Good, I.J. (Ed) *The Scientist Speculates: An Anthology of Partly-Baked Ideas*, New York: Basic Books, pp.192–198.
4. Good, I.J. (1965) Speculations concerning the first ultraintelligent machine, *Advances in Computers* **6**:31–88. This paper was based on talks that Jack gave in 1962 and 1963 and was written in 1963, although it was not published until 1965.
5. For a detailed, neutral discussion of the singularity see Shanahan, M. (2015) *The Technological Singularity*, Cambridge, MA: MIT Press.
6. Interview with Jack Good by P. Husbands and O. Holland, April 2002, Blacksburg, Virginia.
7. Kurzweil, R. (2005) *The Singularity Is Near*, New York: Penguin.
8. Kurzweil, R. (2012) *How to Create a Mind: The Secret of Human Thought Revealed*, New York: Viking Books.
9. Steven Pinker (2008), quote in *IEEE Spectrum* special report on The Singularity, https://spectrum.ieee.org/computing/hardware/tech-luminaries-address-singularity [retrieved Dec 2019].
10. Interview with Jack Good by P. Husbands and O. Holland, April 2002, Blacksburg, Virginia.
11. Husbands, P. (2008) An Interview with John Holland, In Husbands, P., Holland, O., and Wheeler, M. (Eds) *The Mechanical Mind in History*, Cambridge, MA: MIT Press, pp.382–395.
12. Clark, A. (2003) *Natural-Born Cyborgs: Minds, Technologies, and the Future of Human Intelligence*, Oxford: Oxford University Press.

13. For instance, various research products developed by the likes of Johns Hopkins University, but also more commercial products from companies such as Ossur and Blatchford.

14. A new generation of artificial retinas based on 2-D materials, American Chemical Society, August 20, 2018.

15. This is one of Grey Walter's achievements that is much less well known than his development of the first autonomous mobile robots. The demonstration was described in detail in a report in *The People* newspaper on December 17, 1967 (Just think—and your brain will turn the TV off, p.6), as well as being filmed for a contemporary BBC TV documentary.

16. See e.g., Robotic exoskeleton helps a paralyzed man walk at CNET.com [retrieved Jan 2020].

17. You can see Harbisson giving a TED talk on his extra sense at https://www.youtube.com/watch?v=ygRNoieAnzI [retrieved Nov 2020].

## Chapter 7

1. See e.g., https://www.youtube.com/watch?v=cLVCGEmkJs0 [retrieved Jan 2020] for a video of the technology at work.

2. Vincent, J. (2018) Welcome to the automated warehouse of the future, *The Verge May 8*; see https://www.youtube.com/watch?v=4DKrcpa8Z_E [retrieved Jan 2020] for a video of the warehouse in operation.

3. For instance, Covariant, a Californian robotics company, provides such systems which are starting to be installed. See e.g., Ackerman, E. (2020) Covariant uses simple robot and gigantic neural net to automate warehouse picking, *IEEE Spectrum Jan 29*, https://spectrum.ieee.org/automaton/robotics/industrial-robots/covariant-ai-gigantic-neural-network-to-automate-warehouse-picking.

4. Robinson, A., Mulvany, L., and Stringer, D. (2019) Robots take the wheel as autonomous farm machines hit the fields, *Bloomberg May 16*.

5. (2019) UK based Small Robot Company are trialling such machines on various farms, see e.g. *Tucker, I.(2019) Robot farmers, The Observer June 2, p.29..*

6. https://www.bostondynamics.com/spot [retrieved Dec 2019]. Guizzo, E. (2019) Boston Dynamics' Spot robot dog goes on sale, *IEEE Spectrum Sept 24*.

7. https://www.sae.org/standards/content/j3016_201806/ [retrieved Jan 2020].

8. http://rodneybrooks.com/predictions-scorecard-2020-january-01/ [retrieved Jan 2020]. Brooks first made the predictions in 2018 and reviews them every year according to progress or lack of it. His blog makes frequent wise proclamations and observations about robotics and AI.

9. https://www.theverge.com/2019/12/9/21000085/waymo-fully-driverless-car-self-driving-ride-hail-service-phoenix-arizona [retrieved Jan 2020].

10. https://singularityhub.com/2019/08/14/driverless-electric-trucks-are-coming-and-theyll-affect-you-more-than-you-think/ [retrieved Feb 2020].

11. See e.g., http://www.xenex.com/ [retrieved Feb 2020].

12. See e.g., Siripala, T. (2018) Japan's robot revolution in senior care, *The Japan Times, June 9*; for academic studies see e.g., Shibata, T. and Wada, K. (2011) Robot therapy: a new approach for mental healthcare of the elderly—a mini-review, *Gerontology* **57**:378–386 and Abdi, J., Al-Hindawi, A., Ng, T., and Vizcaychipi, M.P. (2018) Scoping review on the use of socially assistive robot technology in elderly care, *BMJ Open* **8**:e018815.

13. Robins, B. and Dautenhahn, K. (2017) The iterative development of the humanoid robot Kaspar: an assistive robot for children with autism, In *Social Robotics: 9th International Conference.*

14. Alcorn, A.M. et al. (2019) Educators' views on using humanoid robots with autistic learners in special education settings in England, *Frontiers in Robotics and AI* **6**:107.

15. Belpaeme, T., Kennedy, J., Ramachandran, A., Scassellati, B., and Tanaka, F. (2018) Social robots for education: a review, *Science Robotics* **3**(21):eaat5954.

16. Originally produced by iRobot, then by Endeavor Robotics, now part of Flir.

17. For further details of *The Senster* see Zivanovic, A. (2009) The technologies of Edward Ihnatowicz, In Brown, P., Gere, C., Lambert, N., and Mason, C. (Eds) *White Heat Cold Logic: British Computer Art 1960–1980*, Cambridge, MA: Leonardo Books/MIT Press, pp.95–110. Brown, P. (2008) The mechanisation of art, In Husbands, P., Holland, O., and Wheeler, M. (Eds) *The Mechanical Mind in History*, Cambridge, MA: MIT Press, pp.259–281. Also https://www.youtube.com/watch?v=hoZb5MTKzQc and

https://www.youtube.com/watch?v=wY85GrYGnyw [retrieved Feb 2020] for images and video of the robot.

18. I have seen Alter-3 in real life, but only witnessed *The Senster* on old movie clips, so perhaps the comparison is unfair.

19. See e.g., Brown, P. et al. (2007) The Drawbots, In *Proceedings of MutaMorphosis: Challenging Arts and Sciences*, Prague, https://mutamorphosis.wordpress.com/2009/03/01/the-drawbots/ [retrieved Feb 2020].

20. See http://patricktresset.com/new/ [retrieved Feb 2020] for example images and videos of the robots at work.

21. See https://www.youtube.com/watch?v=_Xbt8lzWxIQ [retrieved Feb 2020] for an illustrated video of Cohen talking about his work.

22. See Boden, M. (2010) *Creativity and Art: Three Roads to Surprise*, Oxford: Oxford University Press for a deeper discussion of this point.

23. Webb, B. (2001) Can robots make good models of biological behaviour? *Behavioural and Brain Sciences* **24**(6):1033–1050.

24. Chang, E., Matloff, L.Y., Stowers, A.K., and Lentink, D. (2020) Soft biohybrid morphing wings with feathers underactuated by wrist and finger motion, *Science Robotics* **5**:eaay1246. Matloff, L.Y. et al. (2020) How flight feathers stick together to form a continuous morphing wing, *Science* **367**:293–297.

25. Granda, J., Donina, L., Dragone, V., Long, D., and Cronin, L. (2018) Controlling an organic synthesis robot with machine learning to search for new reactivity, *Nature* **559**:377–384.

26. There was a special issue of the journal *Science* (vol. 343, issue 6169, Jan 2014) dedicated to the scientific work of the *Curiosity* mission.

27. Fey, C. and Osborne, M. (2013) *The Future of Employment: How Susceptible Are Jobs to Computerisation?* Oxford: University of Oxford.

28. Frey and Osborne used O*NET, an online service developed for the US Department of Labor.

29. They used a Gaussian mixture model, a popular technique.

30. Notably Arntz, M., Gregory, T., and Zierahn, U. (2016) *The risk of automation for jobs in OECD countries: a comparative analysis*, OECD Social, Employment and Migration Working Papers No. 189; and PwC (2018) *Will robots really steal our jobs? An international analysis of the potential long term impact of automation*, www.pwc.co.uk/

economics [retrieved Feb 2020]; and *The probability of automation in England: 2011 and 2017*, UK Office of National Statistics 2019.

31. E.g., the PwC report referred to in the previous note.

32. Something that some of the tech giants are now starting to do.

33. There are many different proposals for such universal income schemes, for instance Thomas Piketty, author of the massively influential 2014 book on inequality, *Capital in the Twenty-First Century*, Harvard University Press, argued for "a credible and bold basic income" in a blog post on Feb 13, 2017, https://www.lemonde.fr/blog/piketty/2017/02/13/for-a-credible-and-bold-basic-income/ [retrieved Feb 2020].

34. Although, as I write this the 2020 Covid-19 pandemic is sweeping the globe. The eventual effectiveness, or not, of state action, might start to change views on current socio-political and economic structures.

**Chapter 8**

1. Some philosophers think of ethics in wider, or sometimes narrower, terms, but this is the general "common sense" view of ethics that I will be using. For a more detailed discussion see e.g., Crisp, R. (1998) Ethics, In *The Routledge Encyclopedia of Philosophy*, Abingdon, Oxon: Taylor and Francis [retrieved Mar 10, 2020, from https://www.rep.routledge.com/articles/overview/ethics/v-1/sections/ethics-and-meta-ethics, doi:10.4324/9780415249126-L132-1].

2. For a detailed overview of robot ethics and related areas see Veruggio, G., Operto, F., and Bekey, G. (2016) Roboethics: social and ethical implications of robotics, In Siciliano, B. and Khatib, O. (Eds) *Springer Handbook of Robotics, 2nd edn*, Berlin: Springer, pp.2135–2160.

3. Wakabayashi, D. (2018) Self-driving Uber car kills pedestrian in Arizona, where robots roam, *The New York Times March 19*. Elaine Herzberg was pushing a bicycle across a four-lane road in Temple, Arizona, when she was struck by an Uber test vehicle. The car was in autonomous mode with a human safety driver on-board. The driver was not able to react in time. An ensuing investigation revealed some serious potential issues with the vehicle control systems and the way they interacted with the safety driver, its sensor coverage, and the safety driver's levels of

attention. For further details see e.g., https://en.wikipedia.org/wiki/Death_of_Elaine_Herzberg [retrieved Oct 2020].

4. Examples of ethics initiatives and related bodies include: AI governance and safety teams at the Future of Humanity Institute, Oxford University; the Leverhulme Centre for the Future of Intelligence, Cambridge University; and in the last two or three years, increasing numbers of research-funded bodies around the world have launched initiatives that emphasize ethics and responsible AI.

5. The story appeared in the magazine *Astounding Science Fiction*, March 1942.

6. Driver, J. (2014) The history of utilitarianism, In Zalta, E.N. (Ed) *The Stanford Encyclopedia of Philosophy*, Winter 2014 Edition, https://plato.stanford.edu/archives/win2014/entries/utilitarianism-history/ [retrieved March 2020].

7. Goodall, N. (2014) Machine ethics and automated vehicles, In Meyer, G. and Beiker, S. (Eds) *Road Vehicle Automation*, Berlin: Springer, pp.93–102.

8. Wallach, W. and Allen, C. (2009) *Moral Machines: Teaching Robots Right from Wrong*, Oxford: Oxford University Press.

9. For further discussion of these thorny issues see e.g., Lin, P., Abney, K., and Bekey, G. (Eds) (2014) *Robot Ethics: The Ethical and Social Implications of Robotics*, Cambridge, MA: MIT Press.

10. Kranzberg, M. (1986) Technology and history: "Kranzberg's laws," *Technology and Culture* **27**(3):544–560.

11. https://epsrc.ukri.org/research/ourportfolio/themes/engineering/activities/principlesofrobotics/ [retrieved March 2020].

12. Freely available from https://ethicsinaction.ieee.org/ [retrieved Oct 2020].

13. For a discussion of the ethical landscape see Winfield, A. (2019) Ethical standards in robotics and AI, *Nature Electronics* **2**:46–48.

14. https://www.icrac.net/about-icrac/ [retrieved March 2020].

15. Zuboff, S. (2019) *The Age of Surveillance Capitalism*, London: Profile Books.

16. King, T., Aggarwal, N., Taddeo, M., and Floridi, M. (2020) Artificial intelligence crime: an interdisciplinary analysis of foreseeable threats and solutions, *Science and Engineering Ethics* **26**:89–120.

17. Nomura, T., Sugimoto, K., Syrdal, D., and Dautenhahn, K. (2012) Social acceptance of humanoid robots in Japan: a survey for development of the Frankenstein syndrome questionnaire, *Proceedings 12th IEEE-RAS International Conference on Humanoid Robots*.

18. For more on the hitchBOT story see http://www.hitchbot.me/ [retrieved March 2020] or e.g., https://www.theguardian.com/ technology/2015/aug/03/hitchbot-hitchhiking-robot-destroyed-philadelphia [retrieved March 2020].

# INDEX

*For the benefit of digital users, indexed terms that span two pages (e.g., 52–53) may, on occasion, appear on only one of those pages.*

Figures are indicated by *f* following the page number